Springer Series in Materials Science

Volume 316

The Springer Series in Materials Science covers the complete spectrum of materials research and technology, including fundamental principles, physical properties, materials theory and design. Recognizing the increasing importance of materials science in future device technologies, the book titles in this series reflect the state-of-the-art in understanding and controlling the structure and properties of all important classes of materials.

More information about this series at http://www.springer.com/series/856

Daniel Werdehausen

Nanocomposites as Next-Generation Optical Materials

Fundamentals, Design and Advanced Applications

Springer

Daniel Werdehausen
Corporate Research & Technology
Carl Zeiss AG
Jena, Thüringen, Germany

ISSN 0933-033X ISSN 2196-2812 (electronic)
Springer Series in Materials Science
ISBN 978-3-030-75683-3 ISBN 978-3-030-75684-0 (eBook)
https://doi.org/10.1007/978-3-030-75684-0

This Springer imprint is published by the registered company Springer Nature Switzerland AG
The registered company address is: Gewerbestrasse 11, 6330 Cham, Switzerland

Preface

There is no doubt that a deep gap between academic research and commercial applications exists. This gulf is often described by drastic metaphors such as "valley of death". While the use of such language initially sounds grim to researchers, who are enthusiastic about their new technology, its appropriateness can be illustrated by measurable facts. For example, 97% of all hardware-based startups fail; a fate that can often not even be prevented by tens to hundreds of millions of dollars of funding [1]. This illustrates that, as researchers, we should always remember that providing a proof of concept for a new technology or method in a lab is only a first relatively small step towards its commercial success. The metaphor I like best in this context is that technologies and methods are "tools in a toolbox". This picture implies that different tools are available and that one should not chose the latest addition to the toolbox, but rather the "tool" that is the best suited for the task at hand. For a technology or method to be successful, it must hence address an unmet need or, at least, provide significant, but possibly only subjective, advantages. In addition, to complicate things even further, sometimes the most successful technology is not the one that offers the best performance but rather the cheapest one or the one that has the best timing or marketing strategy behind it. The combination of these factors leads to a complexity that makes predictions about the long-term success of a new technology highly unreliable at best. It also poses immense challenges for researchers that attempt to push their technology across the "valley of death". This is because they must leave their comfort zone of academic publishing and successfully tackle a wide range of problems from different fields. To successfully navigate through these challenges, they must form hypotheses about both the technology and the application and be willing to constantly reject and adapt these hypotheses.

In 2017, I joined ZEISS Corporate Research & Technology because, after working on fundamental research topics in the field of strongly correlated electron system, my goal was to dedicate my efforts towards bridging the gap between academia and industry. I was lucky to obtain a Marie Skłodowska-Curie fellowship that allowed me to take advantage of the highly stimulating and application-oriented environment at ZEISS, while, at the same time, maintaining a flexibility that can commonly only be found at universities. Thanks to my great colleagues both at ZEISS and the Friedrich Schiller University Jena, who were my academic partners, this was the best environment for my doctoral research I could have hoped for. This book is based on the work I did during this time. It builds on decades of research into nanostructured optical materials with the goal of developing concepts that bring such materials closer to crossing the "valley of death". To this end, the first part of this book is dedicated to the fundamental properties of amorphous nanocomposites. Building on these findings, I then investigate the potential of such materials with a focus on approaches that could allow for enhancing conventional optical systems in the second part. Some of the content of the following chapters has already been presented in peer-reviewed journal articles [2–7], at conferences [8, 9], and patent applications [10, 11]. The goal of this book is to connect all these different pieces of the puzzle, provide further details, and expand my work beyond these publications. However, following my preceding discussion, I, at this point, already want to emphasize that the concepts and findings I develop and present in this book should not be understood as fully developed for an immediate commercialization. The goal of my work rather is to expand the aforementioned toolbox and develop a new set of adapted hypotheses that can be further build upon by researchers and engineers from different fields. Since I am convinced that the full potential of nanostructured optical materials, and other new technologies, can only be unlocked through such immense joint efforts, I hope that this book will stimulate such works in this direction. In this spirit, I invite all readers to build upon, modify, and challenge all concepts I present in the following chapters.

This book would never have been possible without the support from many different people and institutions. First, I am very grateful to the European Union for providing the funding for this work within the NOLOSS project under the Marie Skłodowska-Curie grant agreement No 675745. Second, while working on the content of this book, I collaborated closely with several great researches on this book's different chapters. Therefore, I want to extend my deepest gratitude to all colleagues who directly contributed to the content presented in the following chapters:

- During my stays in Berlin and Karlsruhe some parts of the code that I rely on in Chap. 4 was developed in collaboration with Dr. Xavier Garcia Santiago. This collaboration was also supported by Prof. Carsten Rockstuhl.
- The experimental data presented in Chap. 5 was obtained in the group of Prof. Harald Giessen at the University of Stuttgart. Specifically, all experiments were performed by Ksenia Weber and Michael Schmidt. The materials were synthesized by Dr. Peter König at the INM-Leibniz Institut für Neue Materialien in Saarbrücken.
- Dr. Sven Burger and Dr. Martin Hammerschmidt provided helpful improvements that made the finite element method simulations presented throughout this thesis more efficient. Dr. Martin Hammerschmidt also helped me by running some of my simulations on the supercomputer HLRN at the Zuse Institute Berlin.

- Dr. Sören Schmidt provided me with his implementation of the wave propagation method. For the simulations presented in Chap. 6, I expanded his code into a modular toolbox that can be used to model optical systems that consist of multiple elements.
- Dr. Markus Seesselberg instigated the idea to replace diffractive lenses by purely refractive elements.
- The optical designs for smartphone cameras presented in Chap. 7 were developed in collaboration with my colleague Hans-Jürgen Dobschal. The benchmark design for these systems was extracted out of a patent by my colleague Dr. Vladan Blahnik.

In addition, I am very grateful to everyone else without whom this work would never have been possible:

- Dr. Manuel Decker for being my main point of contact at ZEISS, many productive discussions, and proofreading hundreds of pages of text countless times.
- Dr. Sven Burger for providing me with access to JCMsuite and many stimulating discussions.
- Prof. Thomas Pertsch for beeing my academic supervisor and for providing me with the opportunity to work on this project.
- Dr. Jörg Petschulat for instigating the project and providing me with the opportunity to join ZEISS Corporate Research.
- Dr. Toralf Scharf for hosting me at EPFL during my first months in the project and all his organizational work for the NOLOSS project.
- Dr. Andrea Berner, Dr. Markus Seesselberg, Dr. Thomas Nobis, Holger Münz, and Dr. Vladan Blahnik for many discussions about optical design topics.
- Dr. Christoph Husemann and Dr. Tanja Teuber for their support within ZEISS Corporate Research.
- Prof. Isabelle Staude for providing me with additional academic guidance.
- Andres Ricardo Vega Perez for proofreading this entire book.
- Prof. Thomas Weiss and Prof. Jörg Schlling for reviewing my original thesis.
- Prof. Harald Giessen for his continued support across many years and shaping my career since my early days as a undergraduate student.

I am also grateful to many other people in the optics and photonics community with whom I had stimulating discussions at conferences or other occasions. I look forward to connecting and continue working with all of you.

Finally, and most importantly, I want to thank my girlfriend and future wife Regina Schaufler, my parents Roland and Sabine Werdehausen, and all my friends from the bottom of my heart for their love, support, and patience. This would have never been possible without you!

Jena, Germany Daniel Werdehausen

About this Book

Despite decades of research nanostructured optical materials have not yet been able to enable a new generation of imaging systems. This book introduces long-awaited concepts towards bridging this gap. It investigates in which fundamental regime nanocomposites can be used as bulk optical materials and highlights the immense potential such materials hold for real-world optical elements and systems. It covers the full spectrum from the fundamental properties of optical materials that result from their microscopic structure to detailed application examples in smartphone cameras. This book also provides an in-depth discussion of how new materials can enable broadband flat optics, that is, diffractive optical elements that can enhance high-end broadband imaging systems. Written by an industry expert, this book connects fundamental research and real-world applications. It is an ideal guide both for optical engineers who are working towards integrating new technologies and researchers who are investigating the fundamental properties of optical materials.

Contents

About the Author

Daniel Werdehausen strives to bridge the gulf between academic research and industrial applications. To this end, he currently serves as a Project Manager in the "Corporate Research & Technology" department of ZEISS. Before taking on this role, he completed his doctoral research as Marie-Sklodowska-Curie fellow in a joint project of ZEISS and the Friedrich Schiller University Jena. During his prior education Daniel Werdehausen worked extensively on both fundamental and applied research topics in different renowned institutions including the École polytechnique fédérale de Lausanne, the Max Planck Institute for Solid State Research in Stuttgart, and the University of Sydney. He also was awarded the Arthur-Fischer awards for one of the best combined B.sc. and M.sc. degrees in Physics at the University of Stuttgart. In his free time, Daniel Werdehausen is a sports enthusiast and enjoys reading scientific literature and books from a wide range of different fields.

Acronyms

AOI—Angle of Incidence:
Angle at which a ray or wave is incident on an optical element or system; Measured in respected to the surface normal or the optical axis of an optical system.

CM—Clausius Mossotti equation:
Equation that provides an analytic relationship between a material's refractive index and the dipole polarizability of its microscopic scatterers (e.g. atoms).

DL—Diffractive lens:
A diffractive optical element that has the functionality of a lens.

DOE—Diffractive optical element:
Optical element that relies on diffraction to modify or redirect the incident light.

EG—Échelette-type grating:
One specific type of diffractive optical elements; Has a surface with a characteristic sawtooth profile.

EMT—Effective medium theory:
Theory that allows for predicting the effective refractive indices of composite materials from the properties of their constituents. Throughout this book mostly used to refer to one specific effective medium theory, namely the Maxwell-Garnett-Mie (MGM) effective medium theory.

FEM—Finite element method:
Numerical method for solving differential equations.

FOV—Field of view:
In this book, the field of view of a diffractive optical element is the full range of angles of incidence across which it maintains a high efficiency. For an optical system, the FOV is the full range of angles of incidence that can be observed with the system.

GRIN—Gradient Refractive Index:
A type of optical element in which the refractive index within the volume of the optical element is varied as a function of the spatial position.

MTF—Modulation transfer function:
Function that quantifies the performance of an optical (imaging) system.

MGM—Maxwell-Garnett-Mie effective medium theory:
One specific type of effective medium theory, which uses Mie theory to determine the dipole polarizability of the inclusions.

NA—Numerical aperture:
Dimensionless quantity which quantifies the range of angles across which an optical system can focus or accept light.

PC—Polycarbonate:
One specific type of polymer.

PMMA—Poly(methyl methacrylate):
Another specific polymer.

TEA—Thin element approximation:
Approximation which reduces an optical element to an infinitely thin layer, which achieves an optical functionality by locally changing an incident wave's phase or amplitude.

WPM—Wave propagation method:
Numerical method for simulating the propagation of electromagnetic waves through a given refractive index distribution.

Chapter 1
Introduction

Unknowingly, humanity has been using nanoparticles to modify the optical properties of materials for centuries. This tradition appears to have started in ancient Mesopotamia and Egypt, where chemically synthesized metallic copper and cuprous oxide nanoparticles were used to produce glasses with a prominent red color already around the fourteenth century BCE [1, 2]. Other famous historical materials that attain their unique properties from their nanocomposite nature are Maya Blue Paint [3–6], which is assumed to date back to the sixth century CE, the Roman Lycurgus Cup dating from the fourth century CE [1, 7], and the colorful glass windows in medieval churches [1, 8]. While these materials were discovered using trial and error and nothing was known about their microscopic structure, nanocomposites today have turned into an immensely diverse research field, in which nanometer-sized inclusions are used to deliberately create improved or even novel properties [9–16]. In fact, as is often the case, evolution has by far preceded modern technology and researchers can therefore draw inspiration from a wide range of naturally occurring materials [11, 17–21]. For example, the unique mechanical properties of bone, bamboo, and nacre all result from their intrinsic nanocomposite structure [11, 17, 18]. Specifically, bone contains calcium phosphate domains with a size between 100 nm and 300 nm, and, depending on the size of these domains, a phase transition from ductile to brittle occurs [18].

Nowadays, potential applications for man-made nanocomposites exceed those of natural and historic materials by far and span across an almost intangible broad range. This range includes, but is not limited to, materials for bone implants [22, 23], therapeutic nanocomposites for the treatment and diagnosis of different diseases [24–26], materials supporting tissue regeneration [27], biodegradable packing materials [28], conductive as well as insulating nanocomposites [15, 29–33], and hard as well as flexible materials [29, 34–36]. This immense variety originates from the heterogeneous nature of nanocomposites, which are defined as materials "com-

D. Werdehausen, *Nanocomposites as Next-Generation Optical Materials*, Springer Series in Materials Science 316, https://doi.org/10.1007/978-3-030-75684-0_1

prising multiple different (nongaseous) phase domains in which at least one type of phase domain is a continuous phase" and "at least one of the phases has at least one dimension of the order of nanometers" [37]. This definition illustrates that an infinite variety of nanocomposites exists, since any continuous phase as the "host matrix" can be combined with one or even multiple materials as the nanometer-sized "inclusions". Moreover, in doing so, the inclusions' concentrations, sizes, shapes, and surface chemistry arise as additional degrees of freedom that can be used to tailor the nanocomposites' overall "effective properties" [14]. This possibility to adjust multiple degrees of freedom holds a high potential for the design of new materials. But it also brings along major challenges, since methods must be developed that allow identifying the best suitable compositions. Therefore, a precise understanding of the physical and chemical processes that affect the nanocomposites' effective properties is indispensable to guide the search for outstanding new materials. Ideally, validated and reliable models, which allow predicting the effective properties directly from the composite materials' microscopic composition, are required.

In optics, the historic tradition of using nanocomposites as optical materials has inspired numerous researchers over the last decades and materials with enhanced linear [13, 16, 38–59, 59–63] or nonlinear [13, 16, 33, 63, 63–68] optical properties have been demonstrated. In fact, publications on optical nanocomposites appear in the literature as early as 1978 [44] and since then a great number of publications have emerged in both the scientific [13, 13, 16, 16, 33, 38–40, 52–59, 59–63, 63, 63, 64, 64–68] as well as the patent [44–51] literature. These materials, to just name a few examples, range from optical switches [33], over filters [60] to materials with enhanced third-harmonic generation [68]. However, the focus of this book is on materials in which nanoparticles are added to a host matrix to modify the linear optical properties, more specifically, the refractive index and its dispersion. In this field, research into nanocomposites has so far almost exclusively focused on high-refractive index materials [41–44, 48–59, 62]. Such materials can, for example, be used as encapsulants for LEDs to increase the extraction efficiencies [55]. In this context, the unique advantage of nanocomposites is that dielectric materials with much higher refractive indices than conventional optical materials [69] can be used as the nanoparticles. Randomly dispersing nanoparticles made from such a material in an amorphous host matrix then ideally leads to a material that combines the advantages of its constituents: (1) the optical homogeneity, isotropy, and translational symmetry of the host material with (2) the high refractive index of the nanoparticles. However, despite the high amount of research effort over a time span of several decades, commercially available high-refractive-index nanocomposites are still scarce. The first mass-produced high-refractive-index nanocomposites for optical applications have only appeared recently and were first restricted materials including ZrO_2-nanoparticles [70–72] (supplier: Pixelligent Technologies LLC). Recently, the supplier of these materials expanded its portfolio with nanocomposites based on TiO_2 nanoparticles. These commercial materials are mainly intended to be used as thin coatings [70–72]. Nanocomposites that can be used as replacements for conventional optical polymers or glasses in bulk optical components are not available. Therefore, the question arises whether nanocomposites are suitable and promising materials for such applications at all. In fact, adding nanoparticles into an optically homogeneous material will always lead to a heterogeneity on a length scale that is larger than the atomic or molecular distances within the host matrix itself. Because of the intrinsic connection between heterogeneity and incoherent scattering [73, 74], the fundamental question consequently is whether a composite material with nanometer-sized inclusions can theoretically act as a homogeneous bulk optical medium at all. I here use the term incoherent scattering to refer to the scattering process that removes energy from a coherent beam of light by scattering it into other directions [73–76]. In fact, while the propagation of light in heterogeneous materials that contain large inclusions and scatter light heavily has been extensively investigated [77–82], the transition from homogeneous to heterogeneous materials has not yet been investigated in depth. For the design of bulk optical nanocomposites, determining at what size of the inclusions this transitions occurres is a key issue because ensuring optical homogeneity is essential to keep incoherent scattering at a negligible level. Furthermore, if nanocomposites could indeed act as bulk optical materials, it has never been investigated if such materials would bring along significant benefits for optical systems that justify their increased complexity.

For the design of optical composite materials multiple effective medium theories have been developed that can be used to predict the composite materials' properties from the properties of their constituents [67, 73, 74, 83–98]. But, the applicability and validity of these theories are under debate [73, 84, 86, 89]. In fact, the concept of an effective refractive index for composite materials itself has fundamental limitations. Some of these limitations have been briefly discussed in the literature [73], but a comprehensive analysis is still missing. Addressing these gaps is a key practical issue, since, as aforementioned, accurate effective medium theories are required to find the most promising materials out of an infinite number of possible combinations. Furthermore, ideally, simple rules on how the different degrees of freedom must be tailored to achieve certain optical properties should be developed to reduce the number of compositions that must be investigated.

With regard to three dimensional nanostructured materials, an important distinction must be made at this point: In this work, I focus on nanocomposites, in which nanoparticles are randomly dispersed in a host matrix. In contrast to such random materials, periodic nanostructured materials have also been widely investigated. In analogy to solid crystals, these materials are commonly referred to as "photonic crystals", and it has been discussed that some of their properties can also be described by an effective refractive index [88, 92, 95–97, 99–111]. However, just like for electrons in crystals, the eigenmodes of photonic crystals are Bloch modes, whereas the eigenmodes of amorphous materials are plane waves. The properties of bulk photonic crystals are consequently determined by the properties of the propagating Bloch modes and their coupling to plane waves at the interfaces to conventional materials [88, 92, 107, 108]. While this can lead to surprising phenomena, this is also the reason why an effective refractive index can commonly only describe some of the properties of photonic crystals and not all [95, 99, 100, 108]. Therefore, these photonic-crystal-type materials can not be readily used as replacements for conventional optical materials and are not further investigated in this book. In contrast, the eigenmodes of the random nanocomposites investigated in this work ideally remain unperturbed plane waves. If it were possible to reach this regime, in which nanocomposites act as homogeneous optical materials, such nanocomposites could be readily used as replacements for conventional optical materials without any restrictions.

Finally, another class of nanoscructured devices, which have recently attracted immense research efforts for their potential to replace conventional optical elements, are metalenses [112–130]. Metalenses, are two dimensional arrangements of nanometer-sized building blocks that locally modify an incident wave's phase or amplitude such that the functionality of a lens is achieved. While the term metalenses has appeared only recently, this technology is based on the concept of blazed binary gratings, which was developed and comprehensively analyzed by Lalanne and coworkers around the year 2000 [131–138]. In fact, it is straightforward to show that the generalized laws of reflection and refraction, which were introduced to describe the properties of metalenses in 2011 [129], are equivalent to the laws of diffraction [139]. Metalenses can consequently be regarded as one specific type of diffractive lenses. This highlights that the research into metalenses goes hand in hand with the long ongoing efforts to develop flat diffractive optical elements (DOEs), which are suitable for optical systems [46, 140–142, 142, 143, 143–145, 145–162, 162–169]. Other technologies for DOEs, which have been investigated in this context, include échelette-type gratings [46, 140–147], holographic elements [148–150], and multilevel DOEs [140, 151–155, 170]. In optical design studies, it has been shown multiple times that integrating DOEs into optical systems can provide significant size or performance benefits for a wide range of different systems [143, 145, 150, 158–163]. However, despite these immense efforts, the widespread integration of DOEs into broadband imaging systems remains elusive with only a few exceptions, for example, a commerical telephoto lens [145]. One of the main reasons for this discrepancy is that DOEs are characterized by an intrinsic periodicity that leads to the appearance of different diffraction orders. This impedes the use of such elements for broadband imaging applications, since these applications require the suppression of all but one diffraction orders to avoid a loss of contrast, colorful flares, and ghost images [146, 164, 166, 171]. While, for a single wavelength, this can be readily achieved using blazed gratings [140], dispersion generally leads to the appearance of spurious diffraction orders for other wavelengths [146]. Furthermore, imaging applications always require a certain field of view [172], that is, a certain range of angles of incidence. To unlock the full potential of DOEs for optical design, it is consequently necessary to develop devices that maintain a high performance across the full parameter range required for a wide range of optical systems. This still is an open challenge.

1.1 Scope and Structure of This Book

The discussion in the previous section has highlighted that, in the fields of nanocomposites and diffractive optics, currently significant gaps between research and practice exist. On the one hand, large amounts of research have been conducted across a time span of several decades, but, on the other hand, the transfer of the scientific concepts into real-world applications is still limited. This work is dedicated to making a contribution towards bridging these gaps. My goal with this book is not to provide fully developed solutions, which are ready to be integrated into commercial products, but rather to develop concepts that guide the way towards unlocking the full potential of the different technologies. At the same time, my goal is to critically evaluate the physical limitations of these technologies. Lastly, I aim to provide a comprehensive analysis of the transition between homogeneous and heterogeneous optical materials, which is a regime of which we currently still have a limited understanding. This goes hand in hand with analyzing the fundamental limitations and the validity of the concept of an effect refractive index. To achieve these different goals, the following chapters are structured around four main research questions:

1. **Can nanocomposites be used as bulk optical materials?** And, if so, how must they be designed? In addition, what are the fundamental limits of the concept of an effective refractive index? These questions are addressed in Chap. 3.
2. **What is the potential of nanocomposites as optical materials?** More specifically, what properties can be achieved? And are effective medium theories accurate tools that can predict their properties? These questions are addressed in Chap. 4.
3. **Do nanocomposites allow for the design of highly efficient diffractive optical elements for broadband applications?** And, if so, are such devices suitable for high-numerical-aperture imaging systems? In addition, can general concepts for how broadband diffractive optical elements must be designed be developed? These questions are addressed in Chap. 5.
4. **Can nanocomposites provide significant benefits for optical systems that outweigh their increased complexity?** And, if so, what are potential applications? These questions are addressed in Chap. 6.

But before tackling these questions I briefly introduce the required fundamental basics and discuss why the development of novel optical materials holds a high potential for improving optical systems in Chap. 2.

References

1. D. Schaming, H. Remita, Nanotechnology: from the ancient time to nowadays. Found. Chem. **17**(3), 187–205 (2015)
2. R.H. Brill, N.D. Cahill, A red opaque glass from sardis and some thoughts on red opaques in general. J. Glass Stud. **30**, 16–27 (1988)
3. R.J. Gettens, Maya blue: an unsolved problem in ancient pigments. Am. Antiquity **27**(4), 557–564 (2017)
4. M.Sánchez del Río, A. Doménech, M.T. Doménech-Carbó, M.L. Vázquez de Agredos Pascual, M. Suárez, E. García-Romero, Chapter 18—the maya blue pigment, in *Developments in Clay Science*, vol. 3, ed by E. Galàn, A. Singer (Elsevier, 2011), pp. 453–481
5. G. Chiari, R. Giustetto, J. Druzik, E. Doehne, G. Ricchiardi, Pre-columbian nanotechnology: reconciling the mysteries of the maya blue pigment. Appl. Phys. A **90**(1), 3–7 (2007)
6. M. Jose-Yacaman, L. Rendon, J. Arenas, M.C. Serra Puche, Maya blue paint: an ancient nanostructured material. Science **273**(5272), 223–225 (1996)
7. I. Freestone, N. Meeks, M. Sax, C. Higgitt, The lycurgus cup—a roman nanotechnology. Gold Bull. **40**(4), 270–277 (2007)
8. J. Delgado, M. Vilarigues, A. Ruivo, V. Corregidor, R.C.d. Silva, L.C. Alves. Characterisation of medieval yellow silver stained glass from Convento de Cristo in Tomar. Portugal. Nuclear Instrum. Methods Phys. Res. Sect. B: Beam Interact. Mater. Atoms **269**(20), 2383–2388 (2011)
9. J.H. Koo, *Fundamentals, Properties, and Applications of Polymer Nanocomposites* (Cambridge University Press, Cambridge, 2017)
10. J. Parameswaranpillai, N. Hameed, T. Kurian, Y. Yu, *Nanocomposite Materials: Synthesis, Properties and Applications* (CRC Press, Boca Raton, 2016)
11. S. Gong, H. Ni, L. Jiang, Q. Cheng, Learning from nature: constructing high performance graphene-based nanocomposites. Mater. Today **20**(4), 210–219 (2017)
12. V.K. Thakur, M.R. Kessler, Self-healing polymer nanocomposite materials: a review. Polymer **69**, 369–383 (2015)
13. D.V. Szabó, T. Hanemann. Polymer nanocomposites for optical applications, in *Advances in Polymer Nanocomposites*, ed. by F. Gao (Woodhead Publishing, 2012), pp. 567–604
14. P.M. Ajayan, L.S. Schadler, P.V. Braun, *Nanocomposite Science and Technology* (Wiley, New York, 2006)
15. N. Grossiord, J. Loos, O. Regev, C.E. Koning, Toolbox for dispersing carbon nanotubes into polymers to get conductive nanocomposites. Chem. Mater. **18**(5), 1089–1099 (2006)
16. L.L. Beecroft, C.K. Ober, Nanocomposite materials for optical applications. Chem. Mater. **9**(6), 1302–1317 (1997)
17. J. Peng, Q. Cheng, High-performance nanocomposites inspired by nature. Adv. Mater. **29**(45) (2017)
18. O.A. Tertuliano, J.R. Greer, The nanocomposite nature of bone drives its strength and damage resistance. Nat. Mater. **15**(11), 1195–1202 (2016)
19. U.G.K. Wegst, H. Bai, E. Saiz, A.P. Tomsia, R.O. Ritchie, Bioinspired structural materials. Nat. Mater. **14**(1), 23–36 (2015)
20. M.A.S. Azizi Samir, F. Alloin, A. Dufresne, Review of recent research into cellulosic whiskers, their properties and their application in nanocomposite field. Biomacromolecules **6**(2), 612–626 (2005)
21. C.B. Thompson, S. Chatterjee, L.T.J. Korley, Gradient supramolecular interactions and tunable mechanics in polychaete jaw inspired semi-interpenetrating networks. Eur. Polymer J. **116**, 201–209 (2019)
22. M.R. Rogel, H. Qiu, G.A. Ameer, The role of nanocomposites in bone regeneration. J. Mater. Chem. **18**(36), 4233–4241 (2008)
23. M.R. Shirdar, N. Farajpour, R. Shahbazian-Yassar, T. Shokuhfar, Nanocomposite materials in orthopedic applications. Frontiers Chem. Sci. Eng. **13**(1), 1–13 (2019)

24. N. Dhas, K. Parekh, A. Pandey, R. Kudarha, S. Mutalik, T. Mehta, Two dimensional carbon based nanocomposites as multimodal therapeutic and diagnostic platform: A biomedical and toxicological perspective. J. Controlled Release **308**, 130–161 (2019)
25. R.J.B. Pinto, M. Nasirpour, J. Carrola, H. Oliveira, C.S.R. Freire, I.F. Duarte, A.M. Grumezescu, Chapter 9—-Antimicrobial properties and therapeutic applications of silver nanoparticles and nanocomposites, in *Antimicrobial Nanoarchitectonics* (Elsevier, 2017), pp. 223–259
26. P.P.D. Kondiah, Y.E. Choonara, P.J. Kondiah, T. Marimuthu, P. Kumar, L.C. du Toit, G. Modi, V. Pillay, Inamuddin, A.M. Asiri, A. Mohammad, 17—Nanocomposites for therapeutic application in multiple sclerosis, in *Applications of Nanocomposite Materials in Drug Delivery* (Woodhead Publishing, 2018), pp. 391–408
27. G. Sandri, M. Bonferoni, S. Rossi, F. Ferrari, C. Aguzzi, C. Viseras, C. Caramella, 19—Clay minerals for tissue regeneration, repair, and engineering, in *Wound Healing Biomaterials*, ed. by M.S. Ågren (Woodhead Publishing, 2016), pp. 385–402
28. J.-W. Rhim, H.-M. Park, C.-S. Ha, Bio-nanocomposites for food packaging applications. Progress Polymer Sci. **38**(10–11), 1629–1652 (2013)
29. Z. Ling, C.E. Ren, M.Q. Zhao, J. Yang, J.M. Giammarco, J. Qiu, M.W. Barsoum, Y. Gogotsi, Flexible and conductive MXene films and nanocomposites with high capacitance. Proc. Natl. Acad. Sci. U. S. A. **111**(47), 16676–81 (2014)
30. S. Ansari, E.P. Giannelis, Functionalized graphene sheet-Poly(vinylidene fluoride) conductive nanocomposites. J. Polymer Sci. Part B: Polymer Phys. **47**(9), 888–897 (2009)
31. Y. Wang, N. Herron, Nanometer-sized semiconductor clusters: materials synthesis, quantum size effects, and photophysical properties. J. Phys. Chem. **95**(2), 525–532 (1991)
32. T. Tanaka, Dielectric nanocomposites with insulating properties. IEEE Trans. Dielectr. Electr. Insul. **12**(5), 914–928 (2005)
33. K. Wang et al., Bacterially synthesized tellurium nanostructures for broadband ultrafast nonlinear optical applications. Nat. Commun. **10**(1), 3985 (2019)
34. J. Musil, Hard and superhard nanocomposite coatings. Surface Coatings Technol. **125**(1–3), 322–330 (2000)
35. Q. Li, L. Chen, M.R. Gadinski, S. Zhang, G. Zhang, H.U. Li, E. Iagodkine, A. Haque, L.-Q. Chen, T.N. Jackson, Q. Wang, Flexible high-temperature dielectric materials from polymer nanocomposites. Nature **523**(7562), 576–579 (2015)
36. A.J. Crosby, J.-Y. Lee, polymer nanocomposites: the nano effect on mechanical properties. Polymer Rev. **47**(2), 217–229 (2007)
37. G. Bogdanic, A. Kuzmić. Definitions of terms related to polymer blends, composites, and multiphase polymeric materials, VII.1. Kemija u Industriji 58 (2009)
38. G. Garnweitner, L.M. Goldenberg, O.V. Sakhno, M. Antonietti, M. Niederberger, J. Stumpe, Large-scale synthesis of organophilic zirconia nanoparticles and their application in organic-inorganic nanocomposites for efficient volume holography. Small **3**(9), 1626–1632 (2007)
39. A. Chatterjee, D. Chakravorty, Glass-metal nanocomposite synthesis by metal organic route. J. Phys. D: Appl. Phys. **22**(9), 1386–1392 (1989)
40. S. Kubo, A. Diaz, Y. Tang, T.S. Mayer, I.C. Khoo, T.E. Mallouk, Tunability of the refractive index of gold nanoparticle dispersions. Nano Lett. **7**(11), 3418–3423 (2007)
41. J.-G. Liu, M. Ueda, High refractive index polymers: fundamental research and practical applications. J. Mater. Chem. **19**(47), 8907–8919 (2009)
42. H. Liu, X. Zeng, X. Kong, S. Bian, J. Chen, A simple two-step method to fabricate highly transparent ITO/polymer nanocomposite films. Appl. Surface Sci. **258**(22), 8564–8569 (2012)
43. C. Lü, Z. Cui, Y. Wang, Z. Li, C. Guan, B. Yang, J. Shen, Preparation and characterization of ZnS-polymer nanocomposite films with high refractive index. J. Mater. Chem. **13**(9), 2189–2195 (2003)
44. N. Nakashima, High refractive index glass compositions. Patent US4082427 (1978)
45. Y. Tanaka, K. Hosokawa, K. Takeuchi, Optical composite material and optical element. Patent US8637588 (2014)

46. H. Ukuda, Optical material, and, optical element, optical system and laminated diffractive optical element using it. Patent US20050110830 (2005)
47. R. Schnell, W. Beier, M. Winkler-Trudewig, Transparenter duroplastischer Komposit aus organischer Matrix mit nanoskaligen Glaspartikeln, Verfahren zur Herstellung desselben und dessen Verwendung. Patent DE 10 2007 017 651 (2007)
48. S. Monickam, D. Peters, G. Cooper, Z. Chen. Nanocomposite formulations for optical applications. Patent US20180223107 (2018)
49. A. Garito, Y.-L. Hsiao, R. Gao, J. Zhu, B. Thomas, A. Panackal, J. Sharma, R. Gao. Optical polymer nanocomposites. Patent US20030175004 (2003)
50. M. Feuillade, G. Cantagrel. Liquid polymerizable composition comprising mineral nanoparticles and its use to manufacture an optical article. Patent US20150203710 (2015)
51. G. Cooper, W. Xu, Z. Chen, High refractive index nanocomposite layer. Patent US10144842. 2018
52. P.T. Chung, C.T. Yang, S.H. Wang, C.W. Chen, A.S. Chiang, C.-Y. Liu, ZrO2/epoxy nanocomposite for LED encapsulation. Mater. Chem. Phys. **136**(2–3), 868–876 (2012)
53. P. Tao, Y. Li, A. Rungta, A. Viswanath, J. Gao, B.C. Benicewicz, R.W. Siegel, L.S. Schadler, TiO2 nanocomposites with high refractive index and transparency. J. Mater. Chem. **21**(46), 18623–18629 (2011)
54. S. Li, M. Meng Lin, M.S. Toprak, D.K. Kim, M. Muhammed, Nanocomposites of polymer and inorganic nanoparticles for optical and magnetic applications. Nano Rev. **1**(1), 5214 (2010)
55. C. Lü, B. Yang, High refractive index organic-inorganic nanocomposites: design, synthesis and application. J. Mater. Chem. **19**(19), 2884–2901 (2009)
56. S. Lee, H.-J. Shin, S.-M. Yoon, D.K. Yi, J.-Y. Choi, U. Paik, Refractive index engineering of transparent ZrO2-polydimethylsiloxane nanocomposites. J. Mater. Chem. **18**(15), 1751–1755 (2008)
57. J.L.H. Chau, Y.-M. Lin, A.-K. Li, W.-F. Su, K.-S. Chang, S.L.-C. Hsu, T.-L. Li, Transparent high refractive index nanocomposite thin films. Mater. Lett. **61**(14–15), 2908–2910 (2007)
58. H.K. Schmidt, Sol-gel nanocomposites as functional optical materials, in *Sol-Gel Optics II*, ed. by J.D. Mackenzie, vol. 1758 (International Society for Optics and Photonics. SPIE, 1992), pp. 396–402
59. C. Lü, Z. Cui, Z. Li, B. Yang, J. Shen, High refractive index thin films of ZnS/polythiourethane nanocomposites. J. Mater. Chem. **13**(3), 526–530 (2003)
60. R.J. Nussbaumer, W.R. Caseri, P. Smith, T. Tervoort, Polymer-TiO2 nanocomposites: a route towards visually transparent broadband UV filters and high refractive index materials. Macromolecular Mater. Eng. **288**(1), 44–49 (2003)
61. A. Biswas, O.C. Aktas, J. Kanzow, U. Saeed, T. Strunskus, V. Zaporojtchenko, F. Faupel, Polymer-metal optical nanocomposites with tunable particle plasmon resonance prepared by vapor phase co-deposition. Mater. Lett. **58**(9), 1530–1534 (2004)
62. C. Lü, C. Guan, Y. Liu, Y. Cheng, B. Yang, PbS/polymer nanocomposite optical materials with high refractive index. Chem. Mater. **17**(9), 2448–2454 (2005)
63. B. Karthikeyan, M. Anija, C.S. Suchand Sandeep, T.M. Muhammad Nadeer, R. Philip, Optical and nonlinear optical properties of copper nanocomposite glasses annealed near the glass softening temperature. Opt. Commun. **281**(10), 2933–2937 (2008)
64. B. Karthikeyan, M. Anija, R. Philip, In situ synthesis and nonlinear optical properties of Au:Ag nanocomposite polymer films. Appl. Phys. Lett. **88**(5), 053104 (2006)
65. A.A. Scalisi, G. Compagnini, L. D'Urso, O. Puglisi, Nonlinear optical activity in Ag-SiO2 nanocomposite thin films with different silver concentration. Appl. Surface Sci. **226**(1–3), 237–241 (2004)
66. R.W. Boyd, R.J. Gehr, G.L. Fischer, J.E. Sipe, Nonlinear optical properties of nanocomposite materials. Pure Appl. Opt. J. Eur. Opt. Soc. Part A **5**(5), 505–512 (1996)
67. X. Zeng, D. Bergman, P. Hui, D. Stroud, Effective-medium theory for weakly nonlinear composites. Phys. Rev. B **38**(15), 10970 (1988)
68. G. Albrecht, M. Hentschel, S. Kaiser, H. Giessen, Hybrid Organic-Plasmonic Nanoantennas with Enhanced Third-Harmonic Generation. ACS Omega **2**(6), 2577–2582 (2017)

69. P. Hartmann, R. Jedamzik, S. Reichel, B. Schreder, Optical glass and glass ceramic historical aspects and recent developments: a Schott view. Appl. Opt. **49**(16), D157–D176 (2010)
70. Z. Chen, Pixelligent zirconia nano-crystals for OLED applications, in *White Paper* (2014)
71. D. Russel, Stabell, Scaling-up Pixelligent nanocrystal dispersions, in *White Paper* (2016)
72. Z. Chen, J. Wang, Pixelligent internal light extraction layer for OLED lighting, in *White Paper* (2014)
73. C.F. Bohren, Applicability of effective-medium theories to problems of scattering and absorption by nonhomogeneous atmospheric particles. J. Atmospheric Sci. **43**(5), 468–475 (1986)
74. C.F. Bohren, D.R. Huffman, *Absorption and Scattering of Light by Small Particles* (Wiley, New York, 2008)
75. L.L. Foldy, The multiple scattering of waves. I. General theory of isotropic scattering by randomly distributed scatterers. Phys. Rev. **67**(3–4), 107–119 (1945)
76. M. Lax, Multiple scattering of waves. Rev. Modern Phys. **23**(4), 287–310 (1951)
77. M.I. Mishchenko, L. Liu, D.W. Mackowski, B. Cairns, G. Videen, Multiple scattering by random particulate media: exact 3D results. Opt. Exp. **15**(6), 2822–2836 (2007)
78. M.I. Mishchenko, L.D. Travis, T-matrix computations of light scattering by large spheroidal particles. Opt. Commun. **109**(1), 16–21 (1994)
79. M.I. Mishchenko, L.D. Travis, A.A. Lacis, *Scattering, Absorption, and Emission of Light by Small Particles* (Cambridge University Press, Cambridge, 2002)
80. A. Macke, M.I. Mishchenko, K. Muinonen, B.E. Carlson, Scattering of light by large non-spherical particles: ray-tracing approximation versus T-matrix method. Opt. Lett. **20**(19), 1934–1936 (1995)
81. Z. Cui, Y. Han, Q. Xu, Numerical simulation of multiple scattering by random discrete particles illuminated by Gaussian beams. J. Opt. Soc. Am. A **28**(11), 2200–2208 (2011)
82. L. Pattelli, A. Egel, U. Lemmer, D.S. Wiersma, Role of packing density and spatial correlations in strongly scattering 3D systems. Optica **5**(9), 1037–1045 (2018)
83. V.A. Markel, Introduction to the Maxwell Garnett approximation: tutorial. J. Opt. Soc. Am. A **33**(7), 1244–1256 (2016)
84. P. Mallet, C.-A. Guérin, A. Sentenac, Maxwell-Garnett mixing rule in the presence of multiple scattering: derivation and accuracy. Phys. Rev. B **72**(1), 014205 (2005)
85. W.T. Doyle, Optical properties of a suspension of metal spheres. Phys. Rev. B **39**(14), 9852 (1989)
86. R. Ruppin, Evaluation of extended Maxwell-Garnett theories. Opt. Commun. **182**(4), 273–279 (2000)
87. T.C. Choy, *Effective Medium Theory: Principles and Applications*, vol. 165 (Oxford University Press, Oxford, 2015)
88. A. Andryieuski, S. Ha, A.A. Sukhorukov, Y.S. Kivshar, A.V. Lavrinenko, Bloch-mode analysis for retrieving effective parameters of metamaterials. Phys. Rev. B **86**(3) (2012)
89. D. Aspnes, Local-field effects and effective-medium theory: a microscopic perspective. Am. J. Phys. **50**(8), 704–709 (1982)
90. P. Chýlek, V. Srivastava, R.G. Pinnick, R. Wang, Scattering of electromagnetic waves by composite spherical particles: experiment and effective medium approximations. Appl. Opt. **27**(12), 2396–2404 (1988)
91. A. Malasi, R. Kalyanaraman, H. Garcia, From Mie to Fresnel through effective medium approximation with multipole contributions. J. Opt. **16**(6), 065001 (2014)
92. K. Mnasri, A. Khrabustovskyi, M. Plum, C. Rockstuhl, Retrieving effective material parameters of metamaterials characterized by nonlocal constitutive relations. Phys. Rev B **99**(3) (2019)
93. G.A. Niklasson, C. Granqvist, O. Hunderi, Effective medium models for the optical properties of inhomogeneous materials. Appl. Opt. **20**(1), 26–30 (1981)
94. D. Ruan, L. Zhu, X. Jing, Y. Tian, L. Wang, S. Jin, Validity of scalar diffraction theory and effective medium theory for analysis of a blazed grating microstructure at oblique incidence. Appl. Opt. **53**(11), 2357–65 (2014)

95. B.T. Schwartz, R. Piestun, Dynamic properties of photonic crystals and their effective refractive index. J. Opt. Soc. Am. B **22**(9), 2018–2026 (2005)
96. D.R. Smith, S. Schultz, P. Markoš, C.M. Soukoulis, Determination of effective permittivity and permeability of metamaterials from reflection and transmission coefficients. Phys. Rev. B **65**(19) (2002)
97. G. Smith, Effective medium theory and angular dispersion of optical constants in films with oblique columnar structure. Opt. Commun. **71**(5), 279–284 (1989)
98. I. Skryabin, A. Radchik, P. Moses, G. Smith, The consistent application of Maxwell-Garnett effective medium theory to anisotropic composites. Appl. Phys. Lett. **70**(17), 2221–2223 (1997)
99. K. Mnasri, F.Z. Goffi, M. Plum, C. Rockstuhl, Homogenization of wire media with a general purpose nonlocal constitutive relation. J. Opt. Soc. Am. B **36**(8), F99 (2019)
100. M. Notomi, Theory of light propagation in strongly modulated photonic crystals: Refraction-like behavior in the vicinity of the photonic band gap. Phys. Rev. B **62**(16), 10696 (2000)
101. M.V. Rybin, D.S. Filonov, K.B. Samusev, P.A. Belov, Y.S. Kivshar, M.F. Limonov, Phase diagram for the transition from photonic crystals to dielectric metamaterials. Nat. Commun. **6**, 10102 (2015)
102. J.D. Joannopoulos, S.G. Johnson, J.N. Winn, R.D. Meade, *Photonic Crystals: Molding the Flow of Light* (Princeton University Press, Princeton, 2011)
103. P. Moitra, Y. Yang, Z. Anderson, I.I. Kravchenko, D.P. Briggs, J. Valentine, Realization of an all-dielectric zero-index optical metamaterial. Nat. Photon. **7**(10), 791 (2013)
104. T. Paul, C. Menzel, W. Śmigaj, C. Rockstuhl, P. Lalanne, F. Lederer, Reflection and transmission of light at periodic layered metamaterial films. Phys. Rev. B **84**(11) (2011)
105. J. Petschulat, C. Menzel, A. Chipouline, C. Rockstuhl, A. Tünnermann, F. Lederer, T. Pertsch. "Multipole approach to metamaterials". Phys. Rev. A **78**(4), 043811 (2008)
106. M.S. Rill, C. Plet, M. Thiel, I. Staude, G. von Freymann, S. Linden, M. Wegener, Photonic metamaterials by direct laser writing and silver chemical vapour deposition. Nat. Mater. **7**, 543 (2008)
107. C. Rockstuhl, C. Menzel, T. Paul, F. Lederer, Homogenization of metamaterials from a Bloch mode perspective, in *2012 International Conference on Electromagnetics in Advanced Applications* (2012), pp. 178–181
108. C. Rockstuhl, T. Paul, F. Lederer, T. Pertsch, T. Zentgraf, T.P. Meyrath, H. Giessen, Transition from thin-film to bulk properties of metamaterials. Phys. Rev. B **77**(3), 035126 (2008)
109. C. Rockstuhl, T. Zentgraf, H. Guo, N. Liu, C. Etrich, I. Loa, K. Syassen, J. Kuhl, F. Lederer, H. Giessen, Resonances of split-ring resonator metamaterials in the near infrared. Appl. Phys. B **84**(1–2), 219–227 (2006)
110. J. Zhou, T. Koschny, M. Kafesaki, C.M. Soukoulis, Negative refractive index response of weakly and strongly coupled optical metamaterials. Phys. Rev B **80**(3) (2009)
111. G. Shvets, I. Tsukerman, *Plasmonics and Plasmonic Metamaterials: Analysis and Applications*, vol. 4 (World Scientific, 2012)
112. S. Wang et al., A broadband achromatic metalens in the visible. Nat. Nanotechnol. **13**(3), 227–232 (2018)
113. H. Liang, A. Martins, B.-H.V. Borges, J. Zhou, E.R. Martins, J. Li, T.F. Krauss, High performance metalenses: numerical aperture, aberrations, chromaticity, and trade-offs. Optica **6**(12), 1461–1470 (2019)
114. M. Decker, I. Staude, Resonant dielectric nanostructures: a low-loss platform for functional nanophotonics. J. Opt. **18**(10), 103001 (2016)
115. M. Decker, I. Staude, M. Falkner, J. Dominguez, D.N. Neshev, I. Brener, T. Pertsch, Y.S. Kivshar, High-efficiency dielectric Huygens' surfaces. Adv. Opt. Mater. **3**(6), 813–820 (2015)
116. S.J. Byrnes, A. Lenef, F. Aieta, F. Capasso, Designing large, high-efficiency, high-numerical-aperture, transmissive meta-lenses for visible light. Opt. Exp. **24**(5), 5110–5124 (2016)
117. W.T. Chen, A.Y. Zhu, V. Sanjeev, M. Khorasaninejad, Z. Shi, E. Lee, F. Capasso, A broadband achromatic metalens for focusing and imaging in the visible. Nat. Nanotechnol. **13**(3), 220–226 (2018)

118. W.T. Chen, A.Y. Zhu, J. Sisler, Z. Bharwani, F. Capasso, A broadband achromatic polarization-insensitive metalens consisting of anisotropic nanostructures. Nat. Commun. **10**(1), 355 (2019)
119. W.T. Chen, A.Y. Zhu, J. Sisler, Y.W. Huang, K.M.A. Yousef, E. Lee, C.W. Qiu, F. Capasso, Broadband achromatic metasurface-refractive optics. Nano Lett. **18**(12), 7801–7808 (2018)
120. M. Decker, W.T. Chen, T. Nobis, A.Y. Zhu, M. Khorasaninejad, Z. Bharwani, F. Capasso, J. Petschulat, Imaging performance of polarization-insensitive metalenses. ACS Photon. (2019)
121. P. Genevet, F. Capasso, F. Aieta, M. Khorasaninejad, R. Devlin, Recent advances in planar optics: from plasmonic to dielectric metasurfaces. Optica **4**(1), 139 (2017)
122. B. Groever, W.T. Chen, F. Capasso, Meta-lens doublet in the visible region. Nano Lett. **17**(8), 4902–4907 (2017)
123. M. Khorasaninejad, F. Capasso, Metalenses: Versatile multifunctional photonic components. Science **358**(6367) (2017)
124. M. Khorasaninejad, W.T. Chen, R.C. Devlin, J. Oh, A.Y. Zhu, F. Capasso, Metalenses at visible wavelengths: Diffraction-limited focusing and subwavelength resolution imaging. Science **352**(6290), 1190–4 (2016)
125. M. Khorasaninejad, Z. Shi, A.Y. Zhu, W.T. Chen, V. Sanjeev, A. Zaidi, F. Capasso, Achromatic metalens over 60 nm bandwidth in the visible and metalens with reverse chromatic dispersion. Nano Lett. **17**(3), 1819–1824 (2017)
126. M. Khorasaninejad, A.Y. Zhu, C. Roques-Carmes, W.T. Chen, J. Oh, I. Mishra, R.C. Devlin, F. Capasso, Polarization-insensitive metalenses at visible wavelengths. Nano Lett. **16**(11), 7229–7234 (2016)
127. R. Sawant, P. Bhumkar, A.Y. Zhu, P. Ni, F. Capasso, P. Genevet, Mitigating chromatic dispersion with hybrid optical metasurfaces. Adv. Mater. **31**(3), e1805555 (2019)
128. A. She, S. Zhang, S. Shian, D.R. Clarke, F. Capasso, Adaptive metalenses with simultaneous electrical control of focal length, astigmatism, and shift. Sci. Adv. **4**(2), eaap9957 (2018)
129. N. Yu, P. Genevet, M.A. Kats, F. Aieta, J.P. Tetienne, F. Capasso, Z. Gaburro, Light propagation with phase discontinuities: generalized laws of reflection and refraction. Science **334**(6054), 333–7 (2011)
130. A.Y. Zhu, W.-T. Chen, M. Khorasaninejad, J. Oh, A. Zaidi, I. Mishra, R. C. Devlin, F. Capasso, Ultra-compact visible chiral spectrometer with meta-lenses. APL Photon. **2**(3), 036103 (2017)
131. P. Lalanne, Waveguiding in blazed-binary diffractive elements. J. Opt. Soc. Am. A **16**(10), 2517 (1999)
132. P. Lalanne, S. Astilean, P. Chavel, E. Cambril, H. Launois, Blazed binary subwavelength gratings with efficiencies larger than those of conventional échelette gratings. Opt. Lett. **23**(14), 1081 (1998)
133. P. Lalanne, S. Astilean, P. Chavel, E. Cambril, H. Launois, Design and fabrication of blazed binary diffractive elements with sampling periods smaller than the structural cutoff. J. Opt. Soc. Am. A **16**(5), 1143–1156 (1999)
134. P. Lalanne, P. Chavel, Metalenses at visible wavelengths: past, present, perspectives. Laser Photon. Rev. **11**(3), 1600295 (2017)
135. P. Lalanne, J.P. Hugonin, P. Chavel, Optical properties of deep lamellar gratings: a coupled bloch-mode insight. J. Lightwave Technol. **24**(6), 2442–2449 (2006)
136. M.-S. L. Lee, P. Lalanne, J.-C. Rodier, E. Cambril, Wide-field-angle behavior of blazed-binary gratings in the resonance domain. Opt. Lett. **25**(23), 1690 (2000)
137. C. Ribot, M.-S.L. Lee, S. Collin, S. Bansropun, P. Plouhinec, D. Thenot, S. Cassette, B. Loiseaux, P. Lalanne, Broadband and efficient diffraction. Adv. Opt. Mater. **1**(7), 489–493 (2013)
138. C. Sauvan, P. Lalanne, M.-S. L. Lee, Broadband blazing with artificial dielectrics. Opt. Lett. **29**(14), 1593 (2004)
139. S. Larouche, D.R. Smith, Reconciliation of generalized refraction with diffraction theory. Opt. Lett. **37**(12), 2391–3 (2012)
140. C.A. Palmer, E.G. Loewen, *Diffraction Grating Handbook* (Newport Corporation New York, 2005)

141. T. Nakai, Diffractive Optical Element. Patent US6587272 (1999)
142. A.J. Glass, K.J. Weible, A. Schilling, H.P. Herzig, D.R. Lobb, J.W. Goodman, M. Chang, A.H. Guenther, T. Asakura, Achromatization of the diffraction efficiency of diffractive optical elements, in Proceedings SPIE 3749, 18th Congress of the International Commission for Optics, vol. 3749 (1999), pp. 378–379
143. S. Thiele, C. Pruss, A.M. Herkommer, H. Giessen, 3D printed stacked diffractive microlenses. Opt. Exp. **27**(24), 35621 (2019)
144. M.T. Gale, Replication techniques for diffractive optical elements. Microelectron. Eng. **34**(3–4), 321–339 (1997)
145. T. Nakai, H. Ogawa, Research on multi-layer diffractive optical elements and their application to camera lenses, in *Diffractive Optics and Micro-Optics* (Optical Society of America, 2002), DMA2
146. B.H. Kleemann, M. Seesselberg, J. Ruoff, Design concepts for broadband high-efficiency DOEs. J. Eur. Opt. Soc. Rapid Publ. **3** (2008)
147. M. Seesselberg, J. Ruoff, B.H. Kleemann, Diffractive optical element for colour sensor has multiple successive curvatures structure at right angles to extension direction. Patent DE102006007432 (2007)
148. J.M. Trapp, M. Decker, J. Petschulat, T. Pertsch, T.G. Jabbour, Design of a 2 diopter holographic progressive lens. Opt. Exp. **26**(25), 32866–32877 (2018)
149. W.C. Sweatt, Describing holographic optical elements as lenses. J. Opt. Soc. Am. **67**(6), 803 (1977)
150. J.M. Trapp, T.G. Jabbour, G. Kelch, T. Pertsch, M. Decker, Hybrid refractive holographic single vision spectacle lenses. J. Eur. Opt. Soc.-Rapid Publ. **15**(1), 14 (2019)
151. G.J. Swanson, *Binary Optics Technology: The Theory and Design of Multi-Level Diffractive Optical Elements* (Report, Lincoln Laboratory Massachusetts Institute of Technology, 1989)
152. S. Banerji, M. Meem, A. Majumder, F.G. Vasquez, B. Sensale-Rodriguez, R. Menon, Imaging with flat optics: metalenses or diffractive lenses? Opt. **6**(6), 805 (2019)
153. G. Kim, J.A. Dominguez-Caballero, R. Menon, Design and analysis of multi-wavelength diffractive optics. Opt. Exp. **20**(3), 2814–23 (2012)
154. N. Mohammad, M. Meem, X. Wan, R. Menon, Full-color, large area, transmissive holograms enabled by multi-level diffractive optics. Sci. Rep. **7**(1), 5789 (2017)
155. P. Wang, N. Mohammad, R. Menon, Chromatic-aberration-corrected diffractive lenses for ultra-broadband focusing. Sci. Rep. **6**, 21545 (2016)
156. Y. Arieli, S. Noach, S. Ozeri, N. Eisenberg, Design of diffractive optical elements for multiple wavelengths. Appl. Opt. **37**(26), 6174 (1998)
157. Y. Arieli, S. Ozeri, N. Eisenberg, S. Noach, Design of a diffractive optical element for wide spectral bandwidth. Opt. Lett. **23**(11), 823 (1998)
158. J.A. Davison, M.J. Simpson, History and development of the apodized diffractive intraocular lens. J. Cataract Refractive Surgery **32**(5), 849–58 (2006)
159. A.A. Kazemi, B. Kress, T. Starner, B.C. Kress, S. Thibault, A review of head-mounted displays (HMD) technologies and applications for consumer electronics, in *Proceedings SPIE 8720, Photonic Applications for Aerospace, Commercial, and Harsh Environments IV* (2013), p. 87200A
160. G.I. Greisukh, E.G. Ezhov, A.V. Kalashnikov, S.A. Stepanov, Diffractive-refractive correction units for plastic compact zoom lenses. Appl. Opt. **51**(20), 4597–604 (2012)
161. G.I. Greisukh, E.G. Ezhov, I.A. Levin, S.A. Stepanov, Design of achromatic and apochromatic plastic micro-objectives. Appl. Opt. **49**(23), 4379–84 (2010)
162. G.I. Greisukh, E.G. Ezhov, S.A. Stepanov, Diffractive-refractive hybrid corrector for achroand apochromatic corrections of optical systems. Appl. Opt. **45**(24), 6137 (2006)
163. T. Stone, N. George, Hybrid diffractive-refractive lenses and achromats. Appl. Opt. **27**(14), 2960–71 (1988)
164. D.A. Buralli, G.M. Morris, Effects of diffraction efficiency on the modulation transfer function of diffractive lenses. Appl. Opt. **31**(22), 4389–96 (1992)

165. D. Faklis, G.M. Morris, Spectral properties of multiorder diffractive lenses. Appl. Opt. **34**(14), 2462–2468 (1995)
166. C. Londono, P.P. Clark, Modeling diffraction efficiency effects when designing hybrid diffractive lens systems. Appl. Opt. **31**(13), 2248–52 (1992)
167. M.D. Missig, G.M. Morris, Diffractive optics applied to eyepiece design. Appl. Opt. **34**(14), 2452–61 (1995)
168. E. Noponen, J. Turunen, A. Vasara, Parametric optimization of multilevel diffractive optical elements by electromagnetic theory. Appl. Opt. **31**(28), 5910–2 (1992)
169. D.W. Sweeney, G.E. Sommargren, Harmonic diffractive lenses. Appl. Opt. **34**(14), 2469–2475 (1995)
170. M. Meem, S. Banerji, C. Pies, T. Oberbiermann, A. Majumder, B. Sensale-Rodriguez, R. Menon, Large-area, high-numerical-aperture multi-level diffractive lens via inverse design. Optica **7**(3), 252 (2020)
171. S. Schmidt, S. Thiele, A. Herkommer, A. Tunnermann, H. Gross, Rotationally symmetric formulation of the wave propagation method-application to the straylight analysis of diffractive lenses. Opt. Lett. **42**(8), 1612–1615 (2017)
172. H. Gross, W. Singer, M. Totzeck, F. Blechinger, B. Achtner, *Handbook of Optical Systems*, vol. 1 (Wiley-VCH, Berlin, 2005)

Chapter 2
Fundamentals of Effective Materials and Diffractive Optics

In this chapter, I introduce the fundamental concepts and relationships from the different fields, including electrodynamics, optical design, materials science, and diffractive optics, which are central to the following chapters. This chapter is not intended to replace a textbook but should rather serve as a reference, which provides the readers with the relevant basics of fields with which they are not familiar.

2.1 Dispersion of Optical Materials and Chromatic Aberration

Microscopically, the optical properties of all materials are determined by a multibody scattering problem that involves all electromagnetic charges that make up the materials. However, for homogeneous materials, this highly complex multibody scattering problem can be reduced to a small number of macroscopic measures using a spatial averaging procedure [1]. In fact, conventional optical materials, like glasses, are also isotropic and consist of small building blocks, that is, atoms or molecules which can be treated as electric dipole scatterers. This allows for fully capturing the linear optical response of such materials using the permittivity (ϵ) as a single scalar measure [1]. In technical optics, the refractive index ($n = \sqrt{\epsilon}$) is then most commonly used instead of the permittivity because n is directly related to the properties of electromagnetic waves within the material [2].

According to the Kramers-Kronig relations, causality dictates that any material with a non-unity refractive index exhibits both absorption and dispersion [2]. Therefore, any material that refracts light, if placed in vacuum, has a refractive index that depends on the wavelength. For transparent optical materials in the visible spec-

D. Werdehausen, *Nanocomposites as Next-Generation Optical Materials*, Springer Series in Materials Science 316, https://doi.org/10.1007/978-3-030-75684-0_2

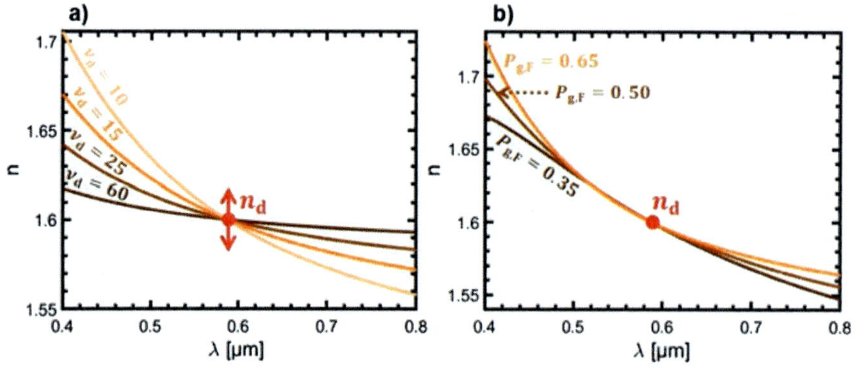

Fig. 2.1 **a** Refractive index (n) as a function of the wavelength (λ) for different Abbe numbers (ν_d). The refractive index at the d-line and the partial dispersion remained fixed at $n_d = 1.6$ and $P_{g,F} = 0.55$, respectively. I used Cauchy's equation (see (2.1)) to obtain the full wavelength dependence of $n(\lambda)$ for the different Abbe numbers. The red arrow visualizes that changing n_d moves the curves up and down. The different lines visualize that lowering ν_d increases the total amount of dispersion. **b** Visualization of the partial dispersion ($P_{g,F}$) for a fixed Abbe number of $\nu_d = 10$. It is evident that $P_{g,F}$ quantifies the curves' steepness on the edges of the spectrum. Reprinted with permission from [3] © The Optical Society

tral range, this wavelength dependence can be accurately described using Cauchy's equation [4]:

$$n(\lambda) = A + \frac{C}{\lambda^2} + \frac{D}{\lambda^4}. \tag{2.1}$$

This equation shows that the refractive indices of transparent materials generally decrease with the wavelength within the visible spectrum. This can be understood by considering that all materials have absorption lines in the ultraviolet spectrum that arise from electronic transitions. Transparency in the visible spectrum can consequently only be achieved if these absorption lines are located at wavelengths sufficiently far below those of the visible spectrum. The Lorentz oscillator model [2] now shows that, if this is the case, the refractive index in the visible spectrum decreases with the wavelength. This also highlights that the magnitude and dispersion of any material's refractive index are fundamentally linked to the number, strength, damping and location of the material's optical resonances.

In technical optics, a material's refractive index and dispersion are usually quantified using the refractive index at the d-line ($n_d = n(\lambda_d = 587.56\,\text{nm})$), the Abbe number ($\nu_d$), and the partial dispersion ($P_{g,F}$):

$$\nu_d = \frac{n_d - 1}{n_F - n_C} \quad \text{and} \quad P_{g,F} = \frac{n_g - n_F}{n_F - n_C}, \tag{2.2}$$

where the subscripts d, F, C, and g correspond to the Fraunhofer lines ($\lambda_g = 435.83\,\text{nm}$, $\lambda_F = 486.13\,\text{nm}$, and $\lambda_C = 656.28\,\text{nm}$) [5]. The influence of all these

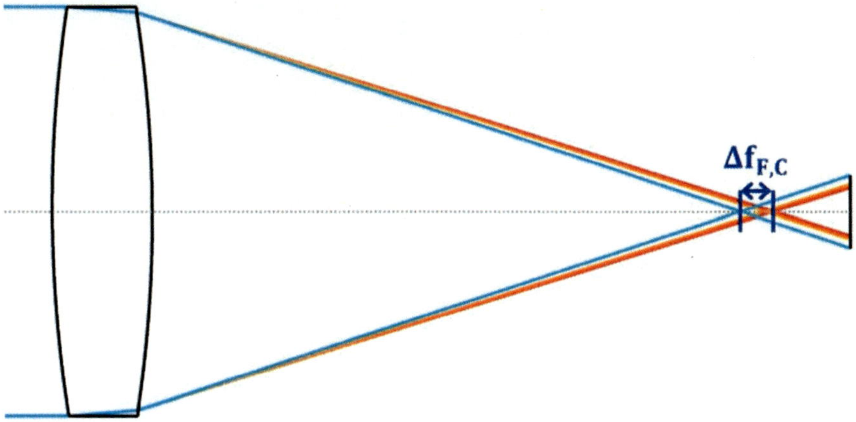

Fig. 2.2 Visualization of the longitudinal chromatic aberration of a single lens. Because of the decrease of the refractive index with increasing wavelength, wavelengths in the blue part of the visible spectrum are focused at a shorter distance from the lens than wavelengths in the red part. $\Delta f_{F,C} = f_F - f_C$ denotes the difference in focal lengths between $\lambda_F = 486.13$ nm, and $\lambda_C = 656.28$ nm

quantities on $n(\lambda)$ is visualized in Fig. 2.1. For producing this figure, I used Cauchy's equation (see (2.1)) to obtain the full wavelength dependence of $n(\lambda)$ for different values of ν_d and $P_{g,F}$. First, the arrow in Fig. 2.1a highlights that modifying n_d shifts $n(\lambda)$ up and down while maintaining its shape. In contrast, lowering the Abbe number increases the refractive index difference between λ_F and λ_C and consequently the total amount of dispersion. This is illustrated by the four different dispersion curves in Fig. 2.1a. Finally, the definition of the partial dispersion in (2.2) illustrates that $P_{g,F}$ quantifies the index change on the blue side of the spectrum (between λ_g and λ_F) relative to the total index difference between the blue (λ_F) and the red (λ_C). Accordingly, Fig. 2.1b illustrates that increasing $P_{g,F}$ increases the steepness of $n(\lambda)$ on the blue side of the spectrum.

2.1.1 Chromatic Aberration

The wavelength dependence of $n(\lambda)$ directly causes the angle of refraction at an interface between two different materials to depend on the wavelength according to Snell's law [5]. Therefore, the focal length of a refractive lens, which relies on curved surfaces to redirect the incident light, also depends on the wavelength. This wavelength dependence is visualized in Fig. 2.2. In optical design, this wavelength dependence is commonly called longitudinal or axial chromatic aberration [5]. For a single lens, the magnitude of the focal shift between λ_F and λ_C, which is highlighted in Fig. 2.2, can be readily determined using its material's Abbe number [5]:

$$\Delta f_{\mathrm{F,C}} = f_{\mathrm{F}} - f_{\mathrm{C}} \approx \frac{f(\lambda_{\mathrm{d}})}{\nu_d}. \tag{2.3}$$

This relationship shows that the focal shift of a single lens increases if its material's Abbe number is decreased. However, the Abbe numbers of all optical materials are positive because their refractive indices decrease as a function of the wavelength; therefore, (2.3) also demonstrates that $\Delta F_{\mathrm{F,C}}$ only takes negative values for negative focal lengths. The correction of longitudinal chromatic aberration with refractive elements hence requires both positive and negative lenses. The simplest system, which is corrected for chromatic aberration, consequently is a two-lens system composed of one positive and one negative lens. Such a system that fulfills the condition $f(\lambda_1) = f(\lambda_2)$ for two distinct wavelengths λ_1 and λ_2, is commonly called an achromat. For $\lambda_1 = \lambda_{\mathrm{F}}$ and $\lambda_2 = \lambda_{\mathrm{C}}$, this condition is fulfilled if [5]:

$$\frac{P_1}{\nu_{\mathrm{d},1}} = -\frac{P_2}{\nu_{\mathrm{d},2}}, \tag{2.4}$$

where $P_i = f_i^{-1}$ is the ith lens' refractive power, f_i its focal length, and $\nu_{\mathrm{d},i}$ its material's Abbe number. Equation (2.4) shows that the key to achieving achromatization is using two materials with different Abbe numbers for the two lenses. To fulfill the constraint that the achromat's overall refractive power must remain positive ($P_{\mathrm{acr}} = \frac{1}{f_{\mathrm{acr}}} = P_1 + P_2 > 0$), the material with the higher amount of dispersion (lower ν_{d}) then has to be used for the negative lens and vice versa. This ensures that the negative lens can compensate for the positive lens' chromatic aberrations, while still maintaining a positive focal length for the overall system ($f_{\mathrm{acr}} > 0$). Finally, the individual focal lengths of an achromat's lenses follow from [6]:

$$f_1 = f_{\mathrm{acr}}\left(\frac{\nu_{\mathrm{d},1} - \nu_{\mathrm{d},2}}{\nu_{\mathrm{d},1}}\right), \quad \text{and} \quad f_2 = f_{\mathrm{acr}}\left(\frac{\nu_{\mathrm{d},2} - \nu_{\mathrm{d},1}}{\nu_{\mathrm{d},2}}\right). \tag{2.5}$$

Compared to single lenses, achromats are distinguished by significantly reduced longitudinal chromatic aberration. However, their focal lengths still deviate from f_{acr} for wavelengths other than the two design wavelengths. This remaining focal shift is commonly called the "secondary spectrum" and its correction generally requires a third lens made of a third material [5]. A system that fulfills $f(\lambda_1) = f(\lambda_2) = f(\lambda_3)$ is then usually referred to as an apochromat [5]. The individual focal lengths of the three lenses within an apochromat with $\lambda_1 = \lambda_{\mathrm{F}}$, $\lambda_2 = \lambda_{\mathrm{C}}$, $\lambda_3 = \lambda_{\mathrm{g}}$, and an overall focal length of f_{apo} can be obtained from [6]:

$$f_1 = f_{\mathrm{apo}}\left(\frac{E(\nu_{\mathrm{d},1} - \nu_{\mathrm{d},3})}{(P_{\mathrm{g,F},2} - P_{\mathrm{g,F},3})\nu_{\mathrm{d},1}}\right), \quad f_2 = f_{\mathrm{apo}}\left(\frac{E(\nu_{\mathrm{d},1} - \nu_{\mathrm{d},3})}{(P_{\mathrm{g,F},3} - P_{\mathrm{g,F},1})\nu_{\mathrm{d},2}}\right),$$
$$\text{and} \quad f_3 = f_{\mathrm{apo}}\left(\frac{E(\nu_{\mathrm{d},1} - \nu_{\mathrm{d},3})}{(P_{\mathrm{g,F},3} - P_{\mathrm{g,F},1})\nu_{\mathrm{d},2}}\right), \tag{2.6}$$

where the factor E follows from [6]:

$$E = \frac{\nu_{d,1}(P_{g,F,2} - P_{g,F,3}) + \nu_{d,2}(P_{g,F,3} - P_{g,F,1}) + \nu_{d,3}(P_{g,F,1} - P_{g,F,2})}{\nu_{d,1} - \nu_{d,3}}. \quad (2.7)$$

I build on these expressions in Chaps. 5 and 6. However, at this point, an important finding from (2.6) and (2.7) is that they directly illustrate that the properties and consequently also the performance of optical systems depend heavily on the materials' refractive indices. In fact, this not only holds true for chromatic aberrations, but also for monochromatic aberrations [5]. This term refers to all aberrations that lead to non-perfect focusing even for light of a single wavelength. Monochromatic aberrations are generally caused by the non-linearity of Snell's law and I refer the interested reader to [5] for an in-depth theoretical treatment of the topic. However, for the understanding of Chaps. 5 and 6, it is important to emphasize that there is also a strong interplay between the correction of monochromatic aberrations and the materials' properties. For example, since the monochromatic aberrations of a lens will increase if its refractive power is increased [5], one goal for the selection of the lenses' materials is to minimize the individual lens' refractive power. This discussion highlights that making a wider range of optical properties n_d, ν_d, and $P_{g,F}$ available could have a significant impact on optical design. In fact, so-called special glasses, which have uncommon dispersion properties, are a key, but also expensive, ingredient for the correction of high-end optical systems [5].

2.2 Analytical Modeling of Nanocomposites

Nanocomposites are made of at least two distinct materials that generally have different refractive indices. Theories that allow for predicting the effective refractive indices of such composite materials from the properties of their constituents are called "effective medium theories". Here, I introduce the effective medium theories that are relevant for the modeling of nanocomposites. In Chap. 3, I provide a detailed discussion and analysis of the controversy surrounding their validity.

To model the effective properties of nanocomposites with hundreds of thousands of nanoparticles, it is first necessary to analyze the response of a single nanoparticle. Throughout this book, I focus on spherical nanoparticles because fabrication techniques that allow for a large-scale production of nanoparticles mostly yield close to spherical shapes. But other types of scatterers, including atoms or molecules, can readily be modeled within the same framework [7–12]. In general, spheres are ideal prototype systems for scattering problems because Mie theory [13] provides a fully analytical solution that describes the optical response of spheres of almost arbitrary size. To this end, Mie theory assumes that a sphere with a permittivity of ϵ_{scat} is embedded in a continous host matrix with a permittivity of ϵ_h. Its key result then is that the sphere's response can be written as a multipole expansion, in which the different components are driven by the respective multipole amplitudes of the incident wave. Accordingly, simple expressions for the extinction σ_{ext} and scattering σ_{scat} cross sections can be obtained [7]:

$$\sigma_{\text{ext}}^{\text{Mie}} = \frac{\lambda^2}{2\pi}\sum_{n=0}^{\infty}(2n+1)\text{Re}(a_n + b_n) \quad \text{and} \tag{2.8}$$

$$\sigma_{\text{scat}}^{\text{Mie}} = \frac{\lambda^2}{2\pi}\sum_{n=0}^{\infty}(2n+1)\text{Re}(|a_n|^2 + |b_n|^2), \tag{2.9}$$

where a_n and b_n denote the electric and magnetic Mie coefficients, which characterize the different multipoles (e.g., $n = 1$: dipoles; $n = 2$: quadrupoles). These coefficients themselves are functions of the refractive index ratio $m = \frac{n_{\text{scat}}}{n_{\text{h}}} = \frac{\sqrt{\epsilon_{\text{scat}}}}{\epsilon_{\text{h}}}$ and the size parameter $x = \pi n_{\text{h}}\frac{d_{\text{scat}}}{\lambda}$, where d_{scat} is the sphere's diameter (see [7]). Finally, because of energy conservation, the absorption cross section (σ_{abs}) reads $\sigma_{\text{abs}} = \sigma_{\text{ext}} - \sigma_{\text{scat}}$. To treat non-spherical particles within the same framework, their multipole response must be determined by other means, e.g. numerical simulations [8, 9].

Mie theory describes the properties of single spherical scatterers. In contrast, effective medium theories serve to predict the properties of large ensembles of scatterers. However, to connect the properties of a distribution of scatterers to those of its individual constituents, several simplifying assumptions are necessary. As a natural starting point for the modelling of nanocomposites, I here start from the Clausius–Mossotti (CM) equation, which is one of the best known and oldest effective medium theories [14]. It assumes (1) that each nanoparticle acts as a dipole scatterer with a dipole moment of $\vec{p} = \alpha_{\text{inc}}\vec{E}_{\text{loc}}$, where α_{scat} is the particles dipole polarizability, and (2) that the local field $\vec{E}_{\text{loc}}$ at the particles' positions is, on average, equal to the external field $\langle\vec{E}_{\text{loc}}\rangle = \vec{E}_{\text{ext}}$. Under these assumptions, the composite material's effective permittivity reads [14]:

$$\epsilon_{\text{eff}} = \epsilon_{\text{h}}\frac{1 + \frac{16f}{d_{\text{scat}}^3}\alpha_{\text{scat}}}{1 - \frac{8f}{d_{\text{scat}}^3}\alpha_{\text{scat}}}, \tag{2.10}$$

where f is the scatterers' volume fraction that follows directly from their number density N_{den} as $f = \frac{4}{3}N_{\text{den}}\pi(\frac{d_{\text{scat}}}{2})^3$ for spherical scatterers. Using the CM equation, the effective refractive index can readily be determined from $n_{\text{eff}} = \sqrt{\epsilon_{\text{eff}}}$. I emphasize that the CM equation fully accounts for dipole-dipole interactions, i.e. multiple scattering, as long as the second approximation ($\langle\vec{E}_{\text{loc}}\rangle = \vec{E}_{\text{ext}}$) remains valid. This is the case for perfectly randomly distributed scatterers or if the scatterers are located on a primitive cubic lattice. But, since hard particles with a finite volume can never reach a perfectly random distribution, there are limits to its validity for nanocomposites at high volume fractions [15]. Furthermore, the CM equation cannot account for statistical fluctuations in the particle distribution. In Chap. 3, I analyze in detail how these aspects affects the CM equation's validity.

Finally, to connect the effective refractive index to the scatterers' properties, suitable expressions for the dipole polarizability α_{scat} must be found. For spherical nanoparticles, this can be achieved using Mie theory. To this end, the dipole amplitude of the scattered field must be divided by the dipole amplitude of the exciting wave, which leads to [16]:

$$\alpha_{\text{scat}}^{\text{Mie}} = i \frac{3 \left(\frac{d_{\text{scat}}}{2}\right)^3}{2x^3} a_1, \tag{2.11}$$

where a_1 is the first order (electric dipole) Mie coefficient. The so-called Maxwell-Garnett-Mie (MGM) effective medium theory now follows directly by substituting the polarizability in (2.11) into the CM equation (2.10). In fact, as long as all other Mie coefficients remain negligible, the polarizability given in (2.11) fully characterizes the particles' response. In contrast, the original Maxwell-Garnett effective medium theory follows from these expressions as the lowest order approximation of the full electric dipole polarizability in (2.11). To this end, the Mie coefficient a_1 has to be expanded into a power series [7] and all but the first term must be neglected. This leads to the well-known quasistatic approximation, which is valid in the limit of infinitely small particles:

$$\alpha_{\text{scat}}^{\text{stat}} = \frac{d_{\text{scat}}}{2}^3 \frac{\epsilon_{\text{scat}} - \epsilon_{\text{h}}}{\epsilon_{\text{scat}} + 2\epsilon_{\text{h}}}. \tag{2.12}$$

The main difference between this quasistatic relationship and the expressions from Mie theory (see (2.11)) is that the latter accounts for retardation within the dipole component, that is, the higher order terms in the expansion of a_1. Therefore, it remains valid up to larger particle sizes. Finally, if the magnetic dipole (coefficient (b_1)) contributes to the particles' response, but all higher order components are still negligible, an effective permeability μ_{eff} can be introduced [17]:

$$\mu_{\text{eff}} = \frac{x^3 + 3ifb_1}{x^3 - \frac{3}{2}ifb_1}. \tag{2.13}$$

In principle, this can lead to an effective material, that is, a material with a nonunity permability, even though all its constituents are non-magnetic. In this case, the effective refractive index follows from $n_{\text{eff}} = \sqrt{\epsilon_{\text{eff}}\mu_{\text{eff}}}$. However, I show in Chap. 3 that such nanocomposites, which include nanoparticles that are large enough for the magnetic dipole to play a role, are not suitable for optical (imaging) systems.

2.3 Nanocomposites Synthesis

Figure 2.3 depicts a scheme of an amorphous nanocomposite, which is the key material class I investigate throughout this book. This scheme visualizes that an amorphous nanocomposite consists of a homogeneous host matrix and a large num-

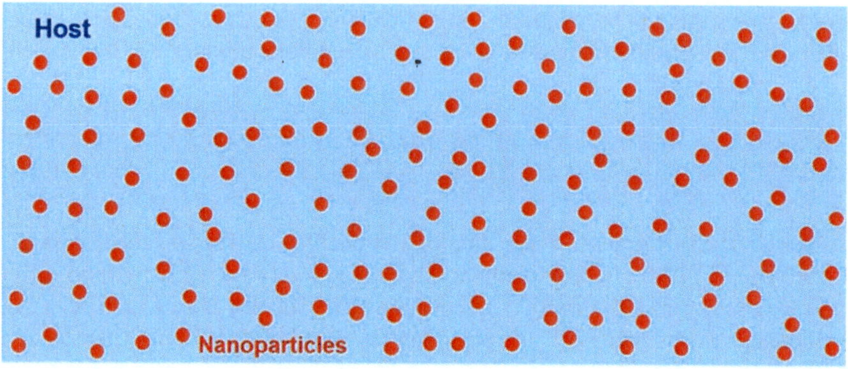

Fig. 2.3 Scheme of an amorphous nanocomposite, which consists of nanoparticles that are randomly dispersed in a homogeneous host matrix. The focus of this book is on materials for which polymers are used as the host matrices

ber of nanoparticles that are randomly dispersed in this host matrix. However, while such a scheme can be readily generated on a computer, realizing an amorphous nanocomposite experimentally can be very challenging. To appreciate these challenges, it is important to understand the microscopic structure of nanocomposites in more detail. First, I here focus on nanocomposites for which polymers are used as the host matrices. In doing so, I generally assume that these polymers can be treated as homogeneous optical materials, that is, that their optical properties can be described by a macroscopic refractive index. However, microscopically, polymers consist of macromolecules, which themselves consist of many repeating subunits [18]. These macromolecules are formed in a process called polymerization. In this process, small molecules, the so-called monomers, undergo reactions through which they become covalently linked to each other and eventually form large interwoven networks or chains [18]. For a polymer to become a homogeneous optical material, it is then essential that the resulting macromolecular structure forms a homogeneous distribution on the scale of the wavelength. In other words, the material must exhibit a constant density throughout the material, that is, density fluctuations on the scale of the wavelength must become negligible. If this condition is not fulfilled, a collimated beam of light that propagates through the polymer experiences incoherent scattering [7]. I discuss and visualize the process of incoherent scattering in more detail in Chap. 4, but, briefly, incoherent scattering causes a collimated beam of light to lose its collimated and directed character by scattering its energy flux into all spatial directions. This, of course, prevents the use of such materials in optical systems, because incoherent scattering leads to stray light that causes a significant loss of contrast in imaging applications. However, modern polymers and manufacturing techniques allow for the fabrication of polymer-based optical components that are almost perfectly homogeneous. Therefore, describing polymers as homogeneous optical materials with a macroscopic refractive index is a highly valid approach.

The addition of nanoparticles to a polymer matrix leads to a second level on which nanocomposites can exhibit a heterogeneity. For the synthesis of nanocomposites that are suitable optical systems, it is hence, first of all, essential that the nanoparticles are well-dispersed in the host matrix. In this context, "well-dispersed" implies that the nanoparticles should ideally be randomly distributed in the host matrix without any biases that lead to deviations from perfectly random distributions. However, fulfilling this condition can be very challenging. This is because the small size and immense surface area to volume ratio causes nanoparticles to exhibit very high surface energies. Therefore, nanoparticles have a high tendency to agglomerate [19]. This can lead to macroscopic phase separation or the formation of larger agglomerates, which strongly scatter light within nanocomposites [19–22]. Therefore, one of the key challenges for the fabrication of optical nanocomposites is to avoid agglomeration by ensuring that the nanoparticles are well dispersed in the host matrix [19, 21]. Since a complete overview over the techniques that have been developed to prevent agglomeration is beyond the scope of this book, I refer the interested reader to the reviews in [19, 23, 24] for further details. However, for nanocomposites based on polymers as the host matrices, which are the materials I focus on in following chapters, the most widely employed approach is to modify the nanoparticles' surfaces with functional molecules, that is, a capping layer. This layer stabilizes the nanoparticles in their environment [19, 22, 23] and, ideally, enables a perfect blending of the nanoparticles into the surrounding molecular structure. For this purpose, an approach called "in situ polymerization" is often used [24]. In this approach, the functionalized nanoparticles are mixed into the monomer before it is polymerized. This, for example, allows for designing the monomers and capping layer such that the capping layer can directly link into the polymer matrix via covalent bonds during polymerization. In fact, I emphasize that the molecular structure of the polymer around the nanoparticles is a key issue for optical nanocomposites. This is because, just like for bulk polymers, fluctuations in the polymer matrix around the nanoparticles can act as defects that scatter light themselves. Therefore, it is necessary to develop carefully designed capping layers for each host matrix. Chemically, this is a highly challenging task and involves carefully optimizing the materials. In fact, the mass-produced ZrO_2-nanoparticles (supplier: Pixelligent Technologies LLC), which are intended for the integration into different polymer matrices, are available with different functionalizations to ensure compatibility with a wide range of polymers [25–28]. However, while, if done correctly, this allows for achieving a very good blending of the nanoparticles in into the host matrix, nanocomposites still possess the intrinsic heterogeneity that is caused by the nanoparticles. Depending on the size of the nanoparticles, this heterogeneity can also lead to large amounts of incoherent scattering. Therefore, I, in Chap. 3, investigate under what fundamental conditions nanocomposites can be used as bulk optical materials.

2.4 Diffractive Optical Elements

One important class of optical elements that I investigate in detail in Chap. 5 are diffractive optical elements (DOEs). In contrast to refractive optical elements, which rely on curved surfaces to redirect the incident light, DOEs are flat and thin elements that shape an incoming wave by varying its phase or amplitude as a function of the transverse position. This definition illustrates that the term DOE includes a wide range of different devices, which span from conventional échelette-type gratings (EGs) [29–36] (Fig. 2.4), over holograms [37–39] and multilevel DOEs [29, 40–44] to metagratings and metalenses [6, 29–37, 39–64].

2.4.1 *Periodic Gratings*

The general properties of DOEs can be best understood using scalar diffraction theory. Furthermore, it is instructive to use the thin element approximation (TEA) and thus reduce the DOE to an infinitely thin plate, which operates by locally changing the incident wave's amplitude or phase. In practice, however, most DOEs operate by changing only the phase to avoid losses. To gain insights into the fundamental properties of such DOEs, I therefore first assume an infinitely periodic grating with a grating period of Λ. Furthermore, I assume a linear phase profile and a maximum phase difference per period, that is, a phase difference between the left and right edge of each period of $\phi_{\max}(\lambda)$. For such a structure, the transmission function for the first period ($x = [0, \Lambda)$) can be written as:

$$t_{\Lambda}^{\mathrm{TEA}}(x, \lambda) = \exp\left[i\phi_{\max}^{\mathrm{TEA}}(\lambda)\frac{x}{\Lambda}\right]. \tag{2.14}$$

The transmission function of the full, infinitely periodic, grating ($T^{\mathrm{TEA}}(x, \lambda)$) now follows directly from (2.14) by periodically repeating $t_{\Lambda}^{\mathrm{TEA}}(x, \lambda)$ into every period of the periodic grating. The spectrum behind the grating can then be calculated as the

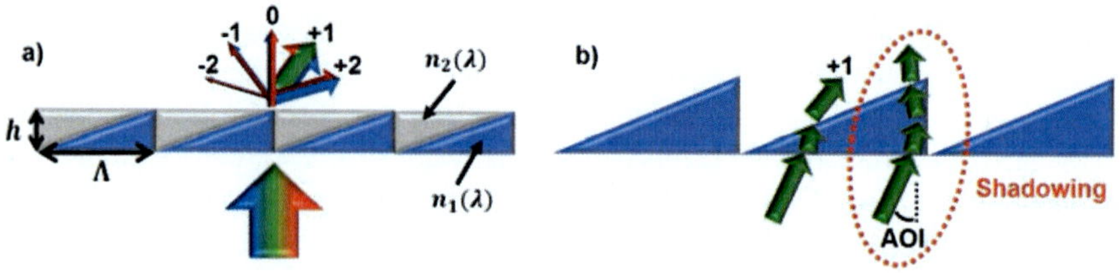

Fig. 2.4 **a** Scheme of an échelette-type grating (EG), which generally disperses broadband light into several diffraction orders. **b** Ray-optical model for shadowing. Shadowing decreases the diffraction efficiency with increasing angles of incidence (AOIs) and decreasing grating periods (Λ). Adapted with permission from [3] © The Optical Society

Fourier transform of $T^{\text{TEA}}(x, \lambda)$. This yields that the spectrum of a periodic grating consists of discrete diffraction orders and directly leads to the well-known grating equation. The grating equation allows for calculating the diffraction angle of the qth order ($\theta_{\text{dif},q}$) as follows [65]:

$$\Lambda \left(\sin(\theta_{\text{AOI}}) - \sin(\theta_{\text{dif},q})\right) = q\lambda, \tag{2.15}$$

where θ_{AOI} is the angle of incidence (AOI). This shows that for a fixed diffraction order and wavelength, the diffraction angle depends only on the direction of the incident wave and the grating period. In addition, for DOEs that only change the local phase and not the amplitude, the power flux in each diffraction order is determined by $\phi_{\max}(\lambda)$. This also directly follows from the Fourier transform of $T(x, \lambda)$. The diffraction efficiency η_q, that is, the percentage of the total transmitted power that propagates in the qth order, can then be obtained from the power fluxes in the different diffraction orders. For perpendicular incidence ($\theta_i = 90°$), this leads to:

$$\eta_q(\lambda) = \text{sinc}^2 \left[\frac{1}{2\pi} (\phi_{\max}(\lambda) - 2\pi q) \right], \tag{2.16}$$

where $\text{sinc(x)} = \frac{\sin(x)}{x}$. This expression demonstrates that, if the condition $\phi_{\max}(\lambda) = 2\pi q$ is fulfilled, all the light will propagate in the qth order. A grating that fulfills this condition for one wavelength and AOI is commonly called a "blazed grating". Specifically, this expression can also be used to determine the efficiencies of échelette-type gratings (EGs; see Fig. 2.4a) within the TEA. To this end, their maximum phase delay per period ($\phi_{\max}(\lambda)$), that is, the phase difference between the left and right edge of each period must be written as:

$$\phi_{\max}^{\text{TEA}}(\lambda) = \frac{2\pi h}{\lambda} [n_2(\lambda) - n_1(\lambda)] = \frac{2\pi h}{\lambda} \Delta n(\lambda), \tag{2.17}$$

where h is the height of the EG and $n_1(\lambda)$ and $n_2(\lambda)$ are the refractive indices of the first and second layer, respectively. As evident from Fig. 2.4a, this expression readily follows from the fact that, on the left edge of each period, a propagating wave only experiences the refractive index $n_2(\lambda)$ and, accordingly, the refractive index $n_1(\lambda)$ determines the phase delay on each period's right edge. Most importantly, (2.17) demonstrates that, for a single wavelength (λ_0), the condition $\phi_{\max}^{\text{TEA}}(\lambda) = 2\pi q$ can be readily fulfilled by choosing the height (h) as follows:

$$h = \frac{q\lambda_0}{\Delta n(\lambda_0)}, \tag{2.18}$$

where λ_0 is then commonly referred to as the grating's design wavelength. However, it can be directly seen from (2.16) and (2.17) that dispersion generally causes the efficiencies to depend strongly on the wavelength. This is one of the main issues that prevents the use of DOEs in broadband applications. Additionally, further problems arise if the EGs' finite heights are taken into account. These problems can be readily understood using the simple ray-optical model illustrated in Fig. 2.4b. This model visualizes that rays, which propagate close the boundaries between the periods, are deflected and hence do not leave the EG in the desired direction. This effect is commonly called shadowing [11, 26]. In fact, it follows from this ray-optical model that the amount of shadowing generally increases with increasing AOIs and heights (h), as well as decreasing grating periods (Λ). Furthermore, the phase profile of an EG, as given in (2.17), only holds for normal incidence. However, the propagation distances of rays that propagate through the EGs increase with increasing AOIs; therefore, the phase delay and hence the efficiency also depend on the AOI, even if the boundaries are neglected. For the sake of conciseness, I from here on use the term "shadowing" as an umbrella term for all these effects. In reality, all shadowing effects, of course, have a wave-optical origin and can therefore only be fully modeled by numerically solving Maxwell's equations [66]. Finally, the fact that the TEA doesn't account for shadowing shows that it allows for obtaining theoretical limits for the efficiencies, which are only achieved in the limit of large grating periods and small AOIs.

The layout of an EG depicted in Fig. 2.4a and its phase profile in (2.17) illustrate that an EG implements a DOE's phase profile by varying the propagation distances in two materials with different refractive indices. In this sense, an EG is one possible "structural embodiment" for a certain functionality that follows from its phase profile. Alternatively, the phase profiles can also be implemented by approximating the linear height profile by discrete levels. This is the approach behind multilevel DOEs [40]. Moreover, the phase profile can be realized by changing the dimensions or orientation of sub-wavelength building blocks. This is the approach used in metagratings and metalenses [48]. The key task for the design of DOEs consequently is to get as close as possible to the ideal functionality that is required for the task at hand. In Chap. 5, I develop approaches for how this can be achieved for broadband optical systems.

2.4.2 *Diffractive Lenses*

While periodic gratings are ideal for studying the fundamental properties of DOEs, broadband imaging systems generally require DOEs that have the functionality of a lens. To achieve the functionality of a spherical lens with a paraxial focal length of $f^{\mathrm{dif}}(\lambda)$, the phase profile must be chosen as follows [6]:

$$\phi(r) = -\frac{\pi}{\lambda f^{\mathrm{dif}}(\lambda)} r^2, \tag{2.19}$$

where r is the radial coordinate. However, there are fundamental limits to the phase delays that can be achieved in thin structures. Therefore, diffractive lenses (DLs) are generally realized by removing unnecessary multiples of 2π from the continous phase profile in (2.19). This leads to the appearance of different segments across

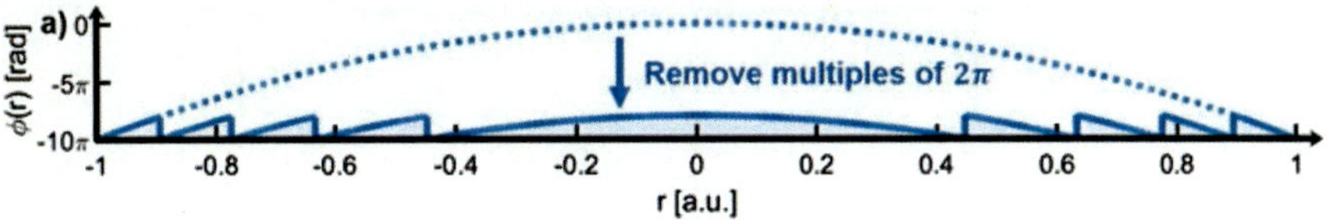

Fig. 2.5 Scheme of a diffractive lens (DL) with a spherical phase profile. The solid line visualizes the periodic structure that is obtained by removing unnecessary multiples of 2π. Adapted from [68]

each of which the phase delay changes between 0 and 2π (Fig. 2.5). It is evident from Fig. 2.5 that focusing is then achieved by decreasing the widths of the segments as the radius is increased. This allows for realizing the functionality of a lens in a thin structure, which only has the thickness required to implement a phase delay of 2π. Therefore, DLs have recently been dubbed "flat optics". But, in analogy to periodic gratings, the downside of introducing discrete steps in the phase profile is that spurious diffraction orders start to appear. For DLs, these spurious orders manifest themselves as higher order foci, which generally lead to stray light and even double images in imaging systems [67]. The key challenge for unlocking the full potential of DLs for broadband optical systems is consequently to design devices that maintain high focusing efficiencies across the full parameter range required for most systems. This raises the question as to which of the different technologies for DOEs is best suited for such applications. I address this question in Chap. 5.

In optical design, DLs are perfectly suited for the correction of chromatic aberrations [6]. To show why this is the case, I now investigate the wavelength dependence of a DL's focal length. Starting from (2.19), I first note that the 2π phase jumps are present for all wavelengths. Therefore, the phase profile of an overall DL, in good approximation, can be assumed to be independent of the wavelength. Substituting this assumption into (2.19) leads to [6]:

$$f^{\mathrm{dif}}(\lambda) = \frac{1}{P^{\mathrm{dif}}(\lambda)} = \frac{\lambda_0}{\lambda} f^{\mathrm{dif}}(\lambda_0). \tag{2.20}$$

This expression shows that the focal length of a DL is inversely proportional to the wavelength. Note that the approximation of a wavelength-independent phase profile neglects the phase differences that arise as a function of the wavelength within each of the DL's segments. From the discussion of infinitely periodic gratings (Sect. 2.4.1), it follows that these changes cause the efficiency to depend on the wavelength. Therefore, they lead to the appearance of higher order foci. But, in good approximation, the wavelength dependent phase changes within each period do not change the location of these foci. I investigate this in more detail in Chap. 5.

The expression for the wavelength dependence of a DL's focal length (2.20) can now be used to quantify a DL's chromatic properties by an effective Abbe number ($\nu_{\mathrm{d}}^{\mathrm{dif}}$) and partial dispersion ($P_{\mathrm{g,F}}^{\mathrm{dif}}$). To this end, the expressions for the refractive power of a DL (2.20) and a refractive lens ($P^{\mathrm{ref}}(\lambda) = (n(\lambda) - 1)\, C_{\mathrm{curv}}$) can be used to rewrite the definitions of ν_{d} and $P_{\mathrm{g,F}}$ as follows [6]:

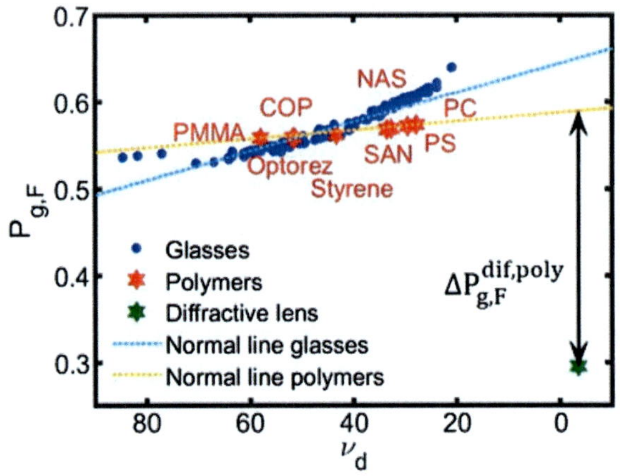

Fig. 2.6 Partial dispersion ($P_{g,F}$) over the Abbe number (ν_d) for the current Schott glasses (blue dots) [69], widely used optical polymers (red stars) [70], and a diffractive lens (DL). The acronyms correspond to: COP—Cyclic olefin copolymer, PMMA—Poly(methyl methacrylate), PC—Polycarbonate, PS—Polystyrene, SAN—Styrol-Acrylnitril-Copolymere. Adapted from [68]

$$\nu_d = \frac{n_d - 1}{n_F - n_C} \underbrace{=}_{P^{ref}(\lambda)} \frac{P(\lambda_d)}{P(\lambda_F) - P(\lambda_C)} \underbrace{=}_{P^{dif}(\lambda)} \frac{\lambda_d}{\lambda_F - \lambda_C} = -3.452 = \nu_d^{dif} \quad \text{and} \tag{2.21}$$

$$P_{g,F} = \frac{n_g - n_F}{n_F - n_C} = \frac{P(\lambda_g) - P(\lambda_F)}{P(\lambda_F) - P(\lambda_C)} = \frac{\lambda_g - \lambda_F}{\lambda_F - \lambda_C} = 0.2956 = P_{g,F}^{dif}, \tag{2.22}$$

where, as highlighted by the curly braces in the first derivation, the expression for the refractive power of a refractive lens was used for the second equality. For the third one, the corresponding relationship for a DL was then used. To compare these quantities to those of conventional optical materials, Fig. 2.6 depicts $P_{g,F}$ as a function of ν_d for conventional optical materials (glasses and polymers) as well as a DL. This so-called partial dispersion diagram demonstrates that a DL is distinguished by highly anomalous values of ν_d^{dif} and $P_{g,F}^{dif}$. Specifically, a DL has a negative effective Abbe number, which cannot be achieved with any transparent optical material (see Sect. 2.1). This, for example, allows for the design of achromats whose two elements both have the same sign of refractive power (see (2.4)). Second, a DL has a highly anomalous $P_{g,F}$ value. As I demonstrate in Chaps. 5 and 6, these properties make DLs unique tools for correcting chromatic aberrations. In Chap. 5, I also show that the anomalous chromatic properties have profound consequences on the properties of DLs in broadband imaging systems.

References

1. G. Russakoff, A derivation of the macroscopic Maxwell equations. Am. J. Phys. **38**(10), 1188–1195 (1970)
2. J.D. Jackson, *Classical Electrodynamics*, 3rd edn. (Wiley, New York, 1999)
3. D. Werdehausen, S. Burger, I. Staude, T. Pertsch, M. Decker, Dispersion-engineered nanocomposites enable achromatic diffractive optical elements. Optica **6**(8), 1031 (2019)
4. J. Mistrik, S. Kasap, H. E. Ruda, C. Koughia, J. Singh, Optical properties of electronic materials: fundamentals and characterization, in *Springer Handbook of Electronic and Photonic Materials*, ed. by P. Kasap Safa Capper (Springer, Cham, 2017), p. 1
5. H. Gross, W. Singer, M. Totzeck, F. Blechinger, B. Achtner, *Handbook of Optical Systems*, vol. 1 (Wiley-VCH, Berlin, 2005)
6. T. Stone, N. George, Hybrid diffractive-refractive lenses and achromats. Appl. Opt. **27**(14), 2960–71 (1988)
7. C.F. Bohren, D.R. Huffman, *Absorption and Scattering of Light by Small Particles* (Wiley, New York, 2008)
8. X.G. Santiago, M. Hammerschmidt, S. Burger, C. Rockstuhl, I. Fernandez-Corbaton, L. Zschiedrich, Decomposition of scattered electromagnetic fields into vector spherical wave functions on surfaces with general shapes. Phys. Rev. B **99**(4), 045406 (2019)
9. I. Fernandez-Corbaton, D. Beutel, C. Rockstuhl, A. Pausch, W. Klopper, Computation of electromagnetic properties of molecular ensembles. Chem. Phys. Chem. **21**, 878 (2020)
10. D. Theobald, A. Egel, G. Gomard, U. Lemmer. Plane-wave coupling formalism for T-matrix simulations of light scattering by nonspherical particles. Phys. Rev. B **96**(3), 033822 (2017)
11. M.I. Mishchenko, L. Liu, D.W. Mackowski, B. Cairns, G. Videen, Multiple scattering by random particulate media: exact 3D results. Opt. Exp. **15**(6), 2822–2836 (2007)
12. M.I. Mishchenko, L.D. Travis, A. Macke, Scattering of light by polydisperse, randomly oriented, finite circular cylinders. Appl. Opt. **35**(24), 4927–4940 (1996)
13. G. Mie, Beiträge zur Optik trüber Medien, speziell kolloidaler Metallösungen. Annalen der Physik **330**(3), 377–445 (1908)
14. V.A. Markel, Introduction to the Maxwell Garnett approximation: tutorial. J. Opt. Soc. Am. A **33**(7), 1244–1256 (2016)
15. P. Mallet, C.-A. Guérin, A. Sentenac. Maxwell-Garnett mixing rule in the presence of multiple scattering: Derivation and accuracy. Phys. Rev. B **72**(1), 014205 (2005)
16. W.T. Doyle, Optical properties of a suspension of metal spheres. Phys. Rev. B **39**(14), 9852 (1989)
17. R. Ruppin, Evaluation of extended Maxwell-Garnett theories. Opt. Commun. **182**(4), 273–279 (2000)
18. J.M.G. Cowie, V. Arrighi, *Polymers: Chemistry and Physics of Modern Materials* (CRC Press, Boca Raton, 2007)
19. S. Li, M. Meng Lin, M.S. Toprak, D.K. Kim, M. Muhammed. Nanocomposites of polymer and inorganic nanoparticles for optical and magnetic applications. Nano Rev. **1**(1), 5214
20. S.K. Kumar, R. Krishnamoorti, Nanocomposites: structure, phase behavior, and properties. Ann. Rev. Chem. Biomolecular Eng. **1**(1), 37–58 (2010)
21. G. Kickelbick, The search of a homogeneously dispersed material-the art of handling the organic polymer/metal oxide interface. J. Sol-Gel Sci. Technol **46**(3), 281–290 (2008)
22. P. Tao, Y. Li, A. Rungta, A. Viswanath, J. Gao, B.C. Benicewicz, R.W. Siegel, L.S. Schadler, TiO2 nanocomposites with high refractive index and transparency. J. Mater. Chem. **21**(46), 18623–18629 (2011)
23. P.H.C. Camargo, K.G. Satyanarayana, F. Wypych, Nanocomposites: synthesis, structure, properties and new application opportunities. Mater. Res. **12**, 1–39 (2009)
24. S.G. Advani, K.-T. Hsaio, *Manufacturing Techniques for Polymer Matrix Composites* (Woodhead Publishing Limited, 2012)
25. Z. Chen, Pixelligent Zirconia Nano-Crystals for OLED applications, in *White Paper* (2014)

26. D. Russel, A. Stabell. Scaling-up pixelligent nanocrystal dispersions, in *White Paper* (2016)
27. Z. Chen, J. Wang, Pixelligent internal light extraction layer for OLED lighting, in *White Paper* (2014)
28. Z. Chen, S. Monickam, M. Weinstein, Low chromatic aberration nanocomposite, in *White Paper* (2015)
29. C.A. Palmer, E.G. Loewen, *Diffraction Grating Handbook* (Newport Corporation New York, 2005)
30. T. Nakai, Diffractive Optical Element. Patent US6587272 (1999)
31. H. Ukuda, Optical material, and, optical element, optical system and laminated diffractive optical element using it. Patent US20050110830 (2005)
32. S. Thiele, C. Pruss, A.M. Herkommer, H. Giessen, 3D printed stacked diffractive microlenses, in Opt. Exp. **27**(24), 35621 (2019)
33. M.T. Gale, Replication techniques for diffractive optical elements. Microelectron. Eng. **34**(3–4), 321–339 (1997)
34. T. Nakai, H. Ogawa, Research on multi-layer diffractive optical elements and their application to camera lenses, in *Diffractive Optics and Micro-Optics. Optical Society of America*, DMA2 (2002)
35. B.H. Kleemann, M. Seesselberg, J. Ruoff, Design concepts for broadband high-efficiency DOEs. J. Eur. Opt. Soc. Rapid Pub. **3** (2008)
36. M. Seesselberg, J. Ruoff, B.H. Kleemann, Diffractive optical element for colour sensor has multiple successive curvatures structure at right angles to extension direction. Patent DE102006007432 (2007)
37. J.M. Trapp, M. Decker, J. Petschulat, T. Pertsch, T.G. Jabbour, Design of a 2 diopter holographic progressive lens. Opt. Exp. **26**(25), 32866–32877 (2018)
38. W.C. Sweatt. Describing holographic optical elements as lenses. J. Opt. Soc. Am. **67**(6), 803 (1977)
39. J.M. Trapp, T.G. Jabbour, G. Kelch, T. Pertsch, M. Decker, Hybrid refractive holographic single vision spectacle lenses. J. Eur. Opt. Soc.-Rapid Pub. **15**(1), 14 (2019)
40. G.J. Swanson, *Binary Optics Technology: The Theory and Design of Multi-Level Diffractive Optical Elements* (Report, Lincoln Laboratory Massachusetts Institute of Technology, 1989)
41. S. Banerji, M. Meem, A. Majumder, F.G. Vasquez, B. Sensale-Rodriguez, R. Menon. Imaging with flat optics: metalenses or diffractive lenses? Opt. **6**(6), 805 (2019)
42. G. Kim, J.A. Dominguez-Caballero, R. Menon, Design and analysis of multi-wavelength diffractive optics. Opt. Exp. **20**(3), 2814–23 (2012)
43. N. Mohammad, M. Meem, X. Wan, R. Menon, Full-color, large area, transmissive holograms enabled by multi-level diffractive optics, Sci. Rep. **7**(1), 5789 (2017)
44. P. Wang, N. Mohammad, R. Menon, Chromatic-aberration-corrected diffractive lenses for ultra-broadband focusing. Sci. Rep. **6**, 21545 (2016)
45. P. Lalanne, Waveguiding in blazed-binary diffractive elements. J. Opt. Soc. Am. A **16**(10), 2517 (1999)
46. P. Lalanne, S. Astilean, P. Chavel, E. Cambril, H. Launois. Blazed binary subwavelength gratings with efficiencies larger than those of conventional échelette gratings, Opt. Lett. **23**(14), 1081 (1998)
47. P. Lalanne, S. Astilean, P. Chavel, E. Cambril, H. Launois, Design and fabrication of blazed binary diffractive elements with sampling periods smaller than the structural cutoff. J. Opt. Soc. Am. A **16**(5), 1143–1156 (1999)
48. P. Lalanne, P. Chavel, Metalenses at visible wavelengths: past, present, perspectives. Laser Photonics Rev. **11**(3), 1600295 (2017)
49. P. Lalanne, J.P. Hugonin, P. Chavel, Optical properties of deep lamellar Gratings: a coupled Bloch-mode insight. J. Lightwave Technol. **24**(6), 2442–2449 (2006)
50. M.-S. L. Lee, P. Lalanne, J.-C. Rodier, E. Cambril, Wide-field-angle behavior of blazed-binary gratings in the resonance domain. Opt. Lett. **25**(23), 1690 (2000)
51. C. Ribot, M.-S.L. Lee, S. Collin, S. Bansropun, P. Plouhinec, D. Thenot, S. Cassette, B. Loiseaux, P. Lalanne, Broadband and efficient diffraction. Adv. Opt. Mater. **1**(7), 489–493 (2013)

52. C. Sauvan, P. Lalanne, M.-S. L. Lee. Broadband blazing with artificial dielectrics. Opt. Lett. **29**(14), 1593 (2004)
53. Y. Arieli, S. Noach, S. Ozeri, N. Eisenberg, Design of diffractive optical elements for multiple wavelengths. Appl. Opt. **37**(26), 6174 (1998)
54. Y. Arieli, S. Ozeri, N. Eisenberg, S. Noach, Design of a diffractive optical element for wide spectral bandwidth. Opt. Lett. **23**(11), 823 (1998)
55. D.A. Buralli, G.M. Morris, Effects of diffraction efficiency on the modulation transfer function of diffractive lenses. Appl. Opt. **31**(22), 4389–96 (1992)
56. J.A. Davison, M.J. Simpson, History and development of the apodized diffractive intraocular lens. J. Cataract Refractive Surgery **32**(5), 849–58 (2006)
57. D. Faklis, G.M. Morris, Spectral properties of multiorder diffractive lenses. Appl. Opt. **34**(14), 2462–2468 (1995)
58. A.J. Glass, K.J. Weible, A. Schilling, H.P. Herzig, D.R. Lobb, J.W. Goodman, M. Chang, A.H. Guenther, T. Asakura, Achromatization of the diffraction efficiency of diffractive optical elements, in *Proceedings SPIE 3749, 18th Congress of the International Commission for Optics*, vol. 3749 (1999), pp. 378–379
59. G.I. Greisukh, E.G. Ezhov, A.V. Kalashnikov, S.A. Stepanov, Diffractive-refractive correction units for plastic compact zoom lenses. Appl. Opt. **51**(20), 4597–604 (2012)
60. G.I. Greisukh, E.G. Ezhov, S.A. Stepanov, Diffractive-refractive hybrid corrector for achroand apochromatic corrections of optical systems. Appl. Opt. **45**(24), 6137 (2006)
61. C. Londono, P.P. Clark, Modeling diffraction efficiency effects when designing hybrid diffractive lens systems. Appl. Opt. **31**(13), 2248–52 (1992)
62. M.D. Missig, G.M. Morris, Diffractive optics applied to eyepiece design. Appl. Opt. **34**(14), 2452–61 (1995)
63. E. Noponen, J. Turunen, A. Vasara, Parametric optimization of multilevel diffractive optical elements by electromagnetic theory. Appl. Opt. **31**(28), 5910–2 (1992)
64. D.W. Sweeney, G.E. Sommargren, Harmonic diffractive lenses. Appl. Opt. **34**(14), 2469–2475 (1995)
65. T.K. Gaylord, M.G. Moharam, Analysis and applications of optical diffraction by gratings. Proc. IEEE **73**(5), 894–937 (1985)
66. O. Sandfuchs, R. Brunner, D. Pätz, S. Sinzinger, J. Ruoff, Rigorous analysis of shadowing effects in blazed transmission gratings. Opt. Lett. **31**(24), 3638 (2006)
67. S. Schmidt, S. Thiele, A. Herkommer, A. Tunnermann, H. Gross, Rotationally symmetric formulation of the wave propagation method-application to the straylight analysis of diffractive lenses. Opt. Lett. **42**(8), 1612–1615 (2017)
68. D. Werdehausen, S. Burger, I. Staude, T. Pertsch, M. Decker, Flat optics in high numerical aperture broadband imaging systems. J. Opt. **22**(6), 065607 (2020)
69. Schott, Optical Glass 2020. Tech. rep. Schott AG (2020)
70. N. Sultanova, S. Kasarova, I. Nikolov, Dispersion properties of optical polymers. Acta Phys. Polonica-Series A General Phys. **116**(4), 585 (2009)

Chapter 3
Design of Bulk Optical Nanocomposites

The concept of a refractive index is a powerful approach that reduces a complex multibody scattering problem, which involves all atoms and molecules that make up the material, to a single measure. Specifically, a scalar refractive index fully quantifies an amorphous material's linear optical response as long as it is made up of building blocks, which are small and dense enough such that they form a homogeneous distribution of electric point dipole scatterers on the scale of the wavelength [1–3]. This holds true for conventional optical materials, e.g. optical glasses, since such materials are composed of atomic and molecular scatterers and are almost perfectly homogeneous [4]. In contrast, as illustrated in Fig. 3.1, materials that contain larger distinct building blocks scatter light heavily and are hence heterogeneous [5–11]. Therefore, the key question for the design of bulk optical nanocomposites, which contain nanometer-sized building blocks by definition, is at what size of the building blocks the materials transition from the homogeneous into the heterogeneous regime. This is also a fundamental research question that has never been investigated in depth.

Obtaining a deeper understanding of the transition between homogeneous and heterogeneous materials requires investigating a multibody scattering problem that involves hundreds of thousands of distinct scatterers (see Fig. 3.1). This is a major challenge, because, on the one hand, the effective medium theories introduced in Sect. 2.2 cannot capture the full complexity of the problem. Specifically, they assume perfectly randomly distributed scatterers and cannot account for statistical fluctuations in the distributions. On the other hand, systematic experimental investigations would entail selectively modifying scatterer distributions that are composed of hundreds of thousands of individual building blocks. This poses a major challenge. Therefore, full wave optical numerical simulations are the best suited tools for gaining deeper insights into the full complexity of the large-scale multibody scattering problem that determines the optical properties of nanocomposites. In fact, since

D. Werdehausen, *Nanocomposites as Next-Generation Optical Materials*, Springer Series in Materials Science 316, https://doi.org/10.1007/978-3-030-75684-0_3

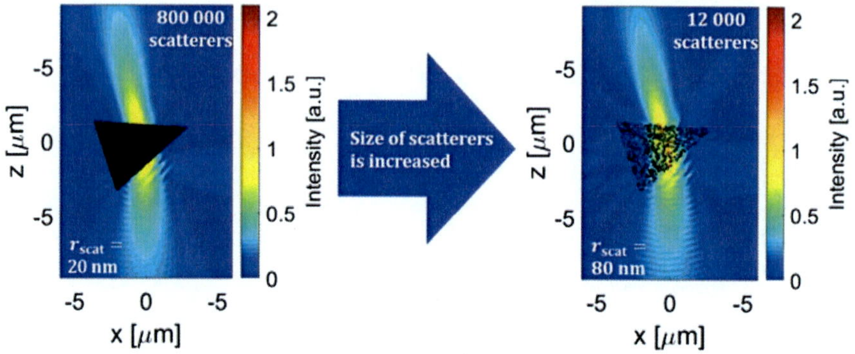

Fig. 3.1 Two dimensional (2D) intensity cross sections obtained from fully three dimensional (3D) wave optical T-matrix simulations. Both panels show a beam of light that propagates through a distribution of nanoparticles (spheres), which are arranged as a 3D prism with a base length in the x, y-plane and a height in the y-direction of 6 μm. The nanoparticles' material is ZrO_2 and the surrounding medium is air. The volume fraction of the nanoparticles is $f = 30\,\%$. The beam of light has a wavelength of $\lambda_0 = 0.8\,\mu\text{m}$, a focus diameter of $\omega_{\text{beam}} = 2\lambda_0 = 1.6\,\mu\text{m}$ and propagates in the positive z-direction. The figure shows that for the smaller nanoparticle radius ($r_{\text{scat}} = 20\,\text{nm}$ - left panel) well-defined transmitted and reflected beams exist. In contrast, for the larger radius ($r_{\text{scat}} = 80\,\text{nm}$ - right panel) the beams are distorted. Reproduced from [14]

molecules or atoms can be treated within the same framework as nanoscopic scatterers [12], nanocomposites are also perfect prototype systems for investigating the properties of optical materials in general. This is because the optical properties of atoms or molecules can be determined from quantum mechanical simulations and then captured in the framework of scattering theory [12]. However, until recently [13], modeling the propagation of light through particle distributions that are composed of hundreds of thousands of individual scatterers was impossible because of the high computational demands of such simulations.

In this chapter, I use large-scale numerical simulations to investigate the transition between the homogeneous and the heterogeneous regimes. To this end, I first introduce a procedure that enables me to obtain reliable refractive index values from numerical simulations of scatterer distributions, which are composed of hundreds of thousands of individual nanoparticles. I then demonstrate that this method indeed enables me to model the regime in which the distribution of scatterers acts as a bulk optical material. This allows me to quantify how bulk optical nanocomposites must be designed and investigate whether the Maxwell-Garnett-Mie effective medium theory (EMT) is an accurate tool for the design of novel nanocomposite materials. Finally, I also show that the concept of an effective refractive index breaks down on multiple level as a material transitions from the homogeneous into the heterogeneous regime. This chapter is based on a recently published paper and investigates the fundamental optical properties of amorphous nanocomposites. Readers who are mainly interested in technical and optical design aspects can quickly skim this chapter.

3.1 Numerical Modelling of Optical Materials

In this section, I introduce my approach for retrieving a material's refractive index from numerical simulations of the underlying multibody scattering problem. As discussed in Sect. 2.2, I focus on nanocomposites that contain nanoparticles with spherical shapes. However, I emphasize that this does not restrict the generality of my approach since both atoms and molecules as well as nanoparticles with other symmetries can be treated using the same framework [12, 15].

3.1.1 Generation of Three Dimensional Particle Distributions

The first step of my procedure, is the generation of three dimensional (3D) nanoparticle distributions. I generate these distributions by placing the nanoparticles into a cuboid that has a length of l_{dist} in the z-direction and a quadratic footprint with a width of w_{dist} in the x/y-plane (see Fig. 3.2a). For generating a distribution, I first chose the target volume fraction (f) and nanoparticle radius (r_{scat}). From the cuboid's volume

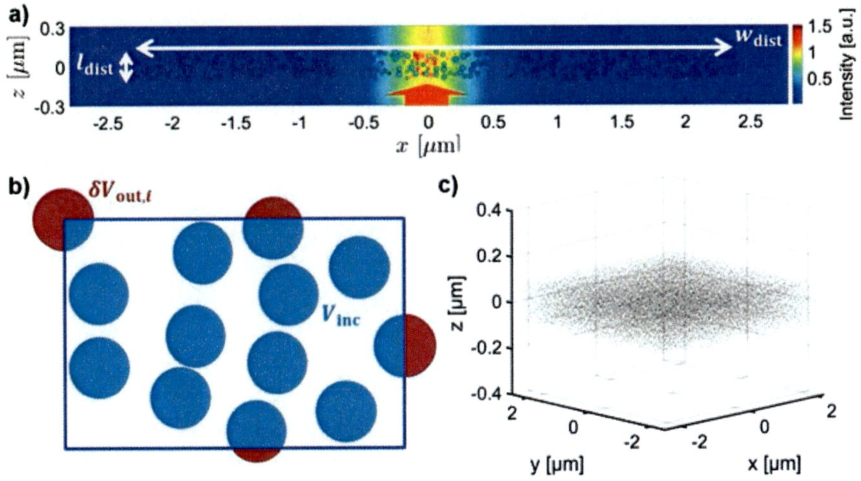

Fig. 3.2 **a** Procedure used for retrieving the effective refractive index from numerical simulations: The nanoparticles (spheres) are placed randomly, but without overlap, into a cuboid with a length of l_{dist} in the z-direction and a quadratic footprint with a width of w_{dist} in the x/y-plane. Subsequently, the distribution is illuminated with a beam of light ($\lambda_0 = 0.8\,\mu\text{m}$) that propagates in the $+z$-direction and the propagation of the beam is modeled using full wave optical simulations. From these simulations, the transmission and reflection coefficients in the normal directions ($k_x = k_y = 0$) are determined. **b** For generating the nanoparticle distributions, the nanoparticles' centers are required to be within the cuboid. The total partial volume outside the cuboid (ΔV_{out}) is then determined and $\lfloor \frac{\Delta V_{\text{out}}}{V_{\text{scat}}} \rfloor$ additional nanoparticles are placed into the cuboid (V_{scat} denotes the volume of a single nanoparticle). **c** Exemplary nanoparticle distribution. Reproduced from [14]

($V_{\text{dist}} = w_{\text{dist}}^2 l_{\text{dist}}$) and that of a single nanoparticle (V_{scat}), I subsequently determine the initial number of nanoparticles (N_0) required to achieve the selected volume fraction ($N_0 = f V_{\text{dist}}/V_{\text{scat}}$). I then successively place the corresponding number of nanoparticles into the cuboid. To this end, I first generate a random position for the center of each new nanoparticle. Subsequently, I check if this new nanoparticle would overlap with any of the previously placed nanoparticles. If there is no overlap, I fix the position of the nanoparticle within the distribution. Otherwise, I generate a new position until there is no overlap. In doing so, I only require the nanoparticles' centers to be inside the cuboid (Fig. 3.2b). If a portion of a nanoparticle's volume is located outside the cuboid, I determine the respective partial volume ($\delta V_{\text{out},i}$) using the analytical expressions derived in [16] (Fig. 3.2b). If the sum over these partial volumes $\left(\Delta V_{\text{out}} = \sum_i \delta V_{\text{out},i}\right)$ exceeds the volume of a single nanoparticle, I place one additional nanoparticle into the cuboid. The overall number of nanoparticles in the cuboid hence is $N_{\text{tot}} = N_0 + \lfloor \frac{\Delta V_{\text{out}}}{V_{\text{scat}}} \rfloor$, where $\lfloor\rfloor$ is the floor operator. In the appendix (Sect. A.1.1), I demonstrate that this approach is essential for achieving well-defined volume fractions within the cuboid and consequently also for retrieving reliable effective refractive index values. An exemplary particle distribution that I generated using this procedure is visualized in Fig. 3.2c. Finally, I note that this procedure can be also applied to other types of building blocks, if their partial volumes outside the cuboid are determined using other techniques, e.g. Monte Carlo methods [17, 18].

3.1.2 Full Wave Optical Simulations

After generating a nanoparticle distribution, I investigate the propagation of a focused beam of light through the distribution (Fig. 3.2a). This approach of using a focused beam whose energy, at the position of the nanoparticle distribution, is almost fully confined within a relatively small volume in space is essential for minimizing the influence of the particle distributions' finite sizes. In fact, below I demonstrate that this allows for retrieving refractive indices that are independent of the distributions' dimensions as long as the distributions are sufficiently large compared to the beam's focus diameter. For all simulations, I model the focused beam as a superposition of plane waves whose amplitudes follow a Gaussian that reaches its maximum for normal incidence. Furthermore, I use a wavelength of $\lambda_0 = 0.8\,\mu\text{m}$ and a focus diameter of $\omega_{\text{beam}} = \lambda_0 = 0.8\,\mu\text{m}$ (see Fig. 3.2a). For each particle distribution, I then perform a full wave optical simulation using the toolbox CELEs. This recently developed toolbox relies on the T-matrix method and exploits the massive parallel computing capabilities of modern GPUs [13]. All simulations were performed on local Linux workstations that are equipped with a "GeForce RTX 2080" graphics card and a CPU with 14 physical cores ("Intel Core i9-7940X"). To ensure that I account for all higher order effects, I include not just the nanoparticles' electric and magnetic dipoles but also their quadrupole responses in all simulations. Furthermore, I use ZrO_2 as the nanoparticles' material, since its high transparency and refractive index in the visible spectral range make it ideally suited for studying the material class of high refractive index dielectric materials [19]. As the homogeneous host matrix, I use PMMA, which is a commonly used optical polymer [20]. Finally, I focus on volume fractions below $f = 30\,\%$, since this is the range that was investigated in most experimental works [21–29].

3.1.3 Retrival Procedure

After the wave optical simulations, I decompose the total field into plane waves. From these data, I then determine the complex transmission (t) and reflection (r) coefficients of the nanoparticle distributions. Because I here focus on random distributions, which are also isotropic, I do so only for the plane waves propagating in the normal directions ($k_x = k_y = 0$). Subsequently, I use three expressions to determine the effective refractive index from the reflection and transmission coefficients. First, I use the expressions for the reflection and transmission coefficients of a slab with a thickness of l_{slab}:

$$r = r_{12} + \frac{t_{12}t_{21}r_{21}\phi^2}{1 - \phi^2 r_{21}^2} \quad \text{and} \quad t = t_{12}t_{21}\frac{\phi}{1 - \phi^2 r_{21}^2}. \tag{3.1}$$

The transmission and reflection coefficients of the two interfaces (r_{ij} and t_{ij} with $i, j \in \{1, 2\}$) in these expressions can be obtained from the Fresnel equations:

$$\begin{aligned} r_{12} &= \frac{n_\text{h} - n^i_\text{eff}}{n_\text{h} + n_\text{eff}}, & t_{12} &= \frac{2n_\text{h}}{n_\text{h} + n^i_\text{eff}}, \\ r_{21} &= \frac{n^i_\text{eff} - n_\text{h}}{n_\text{h} + n^i_\text{eff}}, \quad \text{and} & t_{21} &= \frac{2n^i_\text{eff}}{n_\text{h} + n^i_\text{eff}}, \end{aligned} \tag{3.2}$$

where n_h is the refractive index of the host matrix and n^i_eff with $i \in \{\text{trans, ref}\}$ is the effective refractive index of the nanoparticle distribution that I obtain from the reflected (r) and transmitted (t) components, respectively. Finally, the phase delay ϕ follows from $\phi = \exp(2i\pi n^i_\text{eff} l_\text{slab}\lambda^{-1})$. This approach allows me to determine n^i_eff independently in reflection ($i = \text{ref}$) and transmission ($i = \text{trans}$) by numerically inverting both expressions in (3.1) independently. Furthermore, because the Fresnel equations, as given in (3.2), are only applicable if the nanoparticles' magnetic dipole responses are negligible, I also use an additional equation that relies on both r and t to determine the effective refractive index. This equation holds even for non-unity permeabilities (see Sect. A.1.4 in appendix) and was first derived for photonic-crystal-type metamaterials [30, 31]. Therefore, I here refer to the effective refractive indices obtained from this equation as n^meta_eff.

Finally, for each combination of the macroscopic parameters f, w_dist, l_dist, and r_scat, I retrieve the effective refractive indices of several different random distributions.

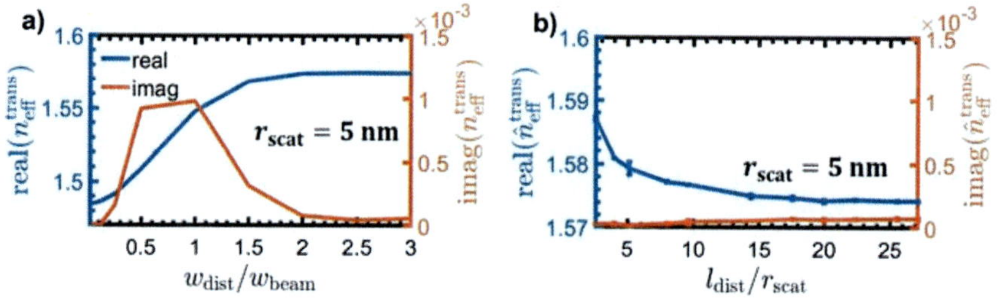

Fig. 3.3 Effective refractive index obtained in transmission ($\hat{n}_{\text{eff}}^{\text{trans}}$) as a function of (**a**) the nanoparticle distribution's width ($w_{\text{dist}}/\omega_{\text{beam}}$) and (**b**) its length ($l_{\text{dist}}/r_{\text{scat}}$). Convergence is achieved for widths exceeding $w_{\text{dist}} = 2.5\omega_{\text{beam}}$ and lengths around $l_{\text{dist}} = 20r_{\text{scat}}$. In both cases a radius of $r_{\text{scat}} = 5$ nm was used. For (**a**), the scatterer distribution with the maximum width was first generated ($w_{\text{dist}} = 3\omega_{\text{beam}}$) and then cropped to smaller widths. For (**b**), different microstates (distributions) were investigated and the depicted data corresponds to the ensemble averages ($\hat{n}_{\text{eff}}^{\text{trans}}$), whereas the error bars denote the standard deviations ($\Delta n_{\text{eff}}^{\text{trans}} = \sigma(\{n_{\text{eff}}^{\text{trans}}\})$). For (**a**), the length (l_{dist}) remained fixed at $l_{\text{dist}} = 20r_{\text{scat}}$, whereas for (**b**), the width remained fixed at $w_{\text{dist}} = 3\omega_{\text{beam}}$ ($\omega_{\text{beam}} = \lambda_0 = 0.8\,\mu$m). The volume fraction remained fixed at $f = 15\,\%$. Reproduced from [14]

This allows me to determine the impact of statistical fluctuations on the distributions' effective refractive indices. Each particle distribution consequently is a microstate within an ensemble that is specified by the macroscopic parameters. I refer to a set of values that I determine for the different microstates within each ensemble as $\{n_{\text{eff}}^i\}$ with $i \in \{\text{trans, ref, meta}\}$. From this data, I then determine the ensemble average $\hat{n}_{\text{eff}}^i = \text{mean}(\{n_{\text{eff}}^i\})$ and the refractive index fluctuations $\Delta n_{\text{eff}}^i = \sigma(\{n_{\text{eff}}^i\})$, where σ corresponds to the standard deviation.

3.2 The Homogeneous Regime—Modelling Bulk Optical Materials at the Single Scatterer Level

Before I investigate how nanocomposites for optical systems must be designed, I here first demonstrate that my approach allows me to numerically model the regime in which a distribution of scatterers acts as a homogeneous bulk optical material.

In general, the bulk regime is reached if the distributions reach a size at which their effective refractive indices become independent of their dimensions. To investigate if it is possible to reach this regime, I consequently first analyze the distributions' effective refractive indices as a function of the distributions' widths (w_{dist}) and lengths (l_{dist}). For this series of simulations, I use a particle radius of only $r_{\text{scat}} = 5$ nm to ensure that the nanoparticles, in good approximation, act as point electric dipole scatterers. For the sake of clarity, I first only investigate the effective refractive index retrieved from the transmitted component ($n_{\text{eff}}^{\text{trans}}$). In the following sections and the appendix, I investigate the differences between $n_{\text{eff}}^{\text{trans}}$, $n_{\text{eff}}^{\text{ref}}$, $n_{\text{eff}}^{\text{meta}}$ and the Maxwell-Garnett-Mie EMT in detail.

Figure 3.3a depicts the effective refractive index obtained in transmission ($n_{\mathrm{eff}}^{\mathrm{trans}}$) as function of the distributions' widths (w_{dist}). For this analysis, I initially generated the nanoparticle distribution with the maximum width ($w_{\mathrm{dist}} = 3\omega_{\mathrm{beam}} = 2.4\,\mu\mathrm{m}$) and subsequently cropped this distribution to smaller widths. Furthermore, I chose a fixed length of $l_{\mathrm{dist}} = 20r_{\mathrm{scat}}$. The corresponding data in Fig. 3.3a demonstrate that both the real and imaginary parts converge to well-defined values for widths exceeding $w_{\mathrm{dist}} = 2.5\omega_{\mathrm{beam}}$. To investigate whether convergence is also achieved with increasing lengths, Fig. 3.3b depicts the effective refractive index obtained in transmission ($\hat{n}_{\mathrm{eff}}^{\mathrm{trans}}$) as a function of l_{dist}. This figure shows that both the real and imaginary parts of $\hat{n}_{\mathrm{eff}}^{\mathrm{trans}}$ converge around $l_{\mathrm{dist}} = 20r_{\mathrm{scat}}$. In the appendix (Sect. A.1.2), I additionally demonstrate that convergence is achieved at this length for a wide range of different radii. In conclusion, the finding that the effective refractive index converges both with increasing w_{dist} and l_{dist} shows that it is indeed possible to reach the bulk regime. Furthermore, Fig. 3.3a demonstrates that the imaginary part of the effective refractive index almost vanishes for sufficiently large widths. This shows that the nanoparticle distribution essentialy acts as a homogeneous lossless optical material. The number of scatterers that was required to reach this regime was around 400 000. Finally, the particle sizes investigated for this analysis exceed the size of simple molecules by only around one order of magnitude [32]. In fact, molecules can be treated within the same framework if their optical response is determined from quantum mechanical simulations [12]. This shows that my approach will soon enable modelling conventional optical materials in the bulk regime at the molecular level.

3.3 The Transition From Homogeneous to Heterogeneous Materials

The possibility of modeling bulk optical materials at the single scatterer level, now enables me to fully analyze the transition from homogeneous to heterogeneous materials. For this analysis, I mainly focus on nanoparticle radii between $r_{\mathrm{scat}} = 10\,\mathrm{nm}$ and $r_{\mathrm{scat}} = 80\,\mathrm{nm}$. Below I demonstrate that this is the region across which the most prominent changes occur for a fixed wavelength of $\lambda_0 = 0.8\,\mu\mathrm{m}$. Moreover, for this range of radii, I keep the distributions' widths and lengths fixed at $w_{\mathrm{dist}} = 6\omega_{\mathrm{beam}}$ and $l_{\mathrm{dist}} = 20r_{\mathrm{scat}}$, respectively. This choice ensures that the distributions' sizes are within the bulk regime, since I increased the width (w_{dist}) by a more than a factor of two compared to the one at which convergence was achieved for small particles (see Fig. 3.3). I did so to compensate for the increase of l_{dist} that is required with increasing radii according to $l_{\mathrm{dist}} = 20r_{\mathrm{scat}}$. Specifically, w_{dist} must be increased for larger values of l_{dist} because, for a focused beam, an increase of the propagation length leads to a more pronounced broadening of the beam. Therefore, a larger value of w_{dist} is required to ensure that the distribution is sufficiently wide compared to the beam. Moreover, in Sect. A.1.2 in the appendix I show that keeping the length

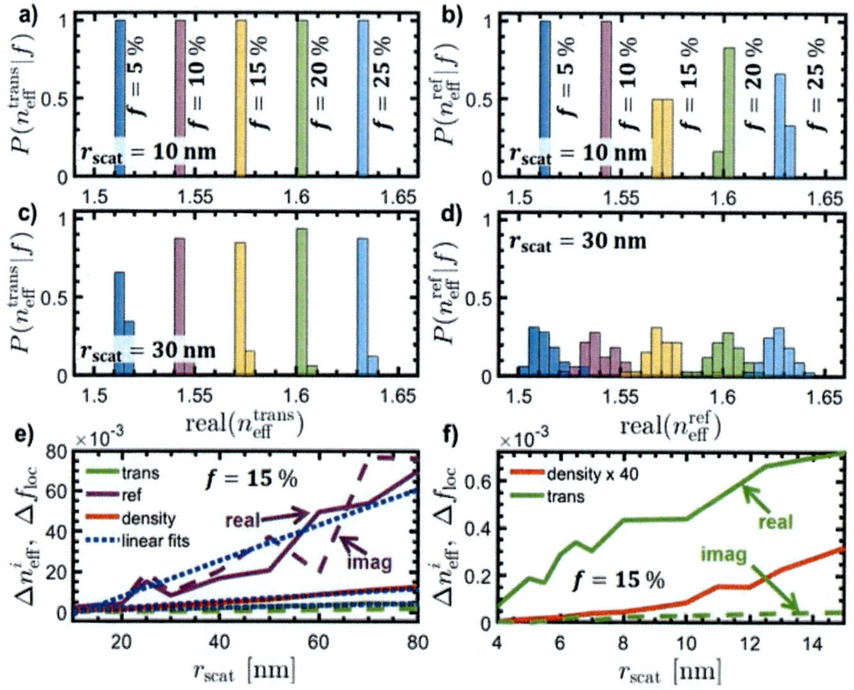

Fig. 3.4 Probability for retrieving a specific value of (**a**, **c**) real(n_{eff}^{trans}) and (**b**, **d**) real(n_{eff}^{ref}) for nanoparticle radii of (a,b) $r_{scat} = 10$ nm and (**c**, **d**) $r_{scat} = 30$ nm at different volume fractions (f). All histograms were generated by analyzing different random nanoparticle distributions for fixed macroscopic parameters (f, w_{dist}, l_{dist}, and r_{scat}). **e** Fluctuations in the effective refractive index (Δn_{eff}^{i}) as determined from the standard deviation ($\Delta n_{eff}^{i} = \sigma(\{n_{eff}^{i}\})$ as a function of r_{scat} within $10\,\text{nm} \leq r_{scat} \leq 80\,\text{nm}$. The red line denotes the local density fluctuations of the nanoparticle distributions ($\Delta f_{loc} = \sigma(\{f_{loc}(\Delta V_i)\})$ on the scale of the wavelength. The dashed blue lines illustrate that both the density fluctuations and the refractive index fluctuations increase close to linearly with r_{scat}. **f** Effective refractive index fluctuations obtained in transmission and local density fluctuations as a function of r_{scat} within $4\,\text{nm} \leq r_{scat} \leq 15\,\text{nm}$. The local density fluctuations were multiplied by a factor of 40 (red line). Reproduced from [14]

fixed at $l_{dist} = 20r_{scat}$ also ensures convergence of $\hat{n}_{eff}^{trans}$ for $r_{scat} > 10$ nm. Finally, to evaluate what radii are required for nanocomposites to be able to replace conventional bulk optical materials, I additionally analyze the range between $r_{scat} = 4$ nm and $r_{scat} = 10$ nm.

3.3.1 Refractive Index Fluctuations

Modern optical materials are distinguished by highly accurate and reproducible refractive indices [4]. In other words, all microstates within an ensemble that is specified by a given set of macroscopic parameters exhibit the same refractive index. In contrast, it can be expected that statistical fluctuations in the nanoparticle distributions start to affect the effective refractive indices outside the homogeneous regime. To analyze this aspect in detail, I generated different random nanoparticle distributions for each ensemble, that is, for each set of the macroscopic parameters f, w_{dist}, l_{dist}, and r_{scat}. First, I did so for two fixed nanoparticle radii of $r_{\text{scat}} = 10\,\text{nm}$ and $r_{\text{scat}} = 30\,\text{nm}$ as well as different volume fractions between $f = 5\,\%$ and $f = 25\,\%$. Thus I obtained the histograms presented in Fig. 3.4a–d. These histograms demonstrate that, at the smaller radius of $r_{\text{scat}} = 10\,\text{nm}$, the effective refractive indices retrieved in transmission ($n_{\text{eff}}^{\text{trans}}$; see Fig. 3.4a) are well-defined, whereas the values retrieved in reflection exhibit small fluctuations ($n_{\text{eff}}^{\text{ref}}$; Fig. 3.4b). Already at $r_{\text{scat}} = 10\,\text{nm}$, which corresponds to a size-to-wavelength ratio of $\frac{d_{\text{scat}}}{\lambda} = \frac{2r_{\text{scat}}}{\lambda} = \frac{1}{40}$, the effective refractive index is consequently not uniquely determined by the macroscopic parameters f, w_{dist}, l_{dist}, and r_{scat}, but also depends on the specific nanoparticle positions of the microstate. Furthermore, Fig. 3.4c, d show that the refractive index fluctuations are drastically enhanced at the larger particle radius of $r_{\text{scat}} = 30\,\text{nm}$. At this radius, the effective refractive indices retrieved in transmission display small fluctuations (Fig. 3.4d) and large fluctuations that span across 0.5 refractive index units are present in reflection (Fig. 3.4d). Combined these findings hence demonstrate that, outside the homogeneous regime, the effective refractive index no longer is a well-defined quantity. It rather critically depends on the specific configuration of the microstate, that is, the scatterer distribution. This shows that the concept of an effective refractive index looses most of its validity outside the homogeneous regime because it no longer describes all microstates within an ensemble.

As the second step, to analyze the influence of the nanoparticle radius on the refractive index fluctuations systematically, I selected a volume fraction of $f = 15\,\%$ and varied the nanoparticle radius between $r_{\text{scat}} = 10\,\text{nm}$ and $r_{\text{scat}} = 80\,\text{nm}$. For each radius, I then quantified the refractive index fluctuations using the standard deviation ($\Delta n_{\text{eff}}^{i} = \sigma(\{n_{\text{eff}}^{i}\})$). Accordingly, Fig. 3.4e depicts the fluctuations in the real and imaginary parts of the effective refractive index over r_{scat}. These data show that the refractive index fluctuations in both reflection and transmission increase close to linearly with r_{scat}. Furthermore, Fig. 3.4e directly confirms that the fluctuations are much larger in reflection than in transmission at all radii. This can most likely partly be explained by the fact that the amplitude of the reflected component is much smaller than that of the transmitted one ($|t| \gg |r|$). Furthermore, it is likely that the reflected component is more sensitive to the distribution's first interface. In the appendix (Sect. A.1.1), I show that this assumption is substantiated by simulations for which I used alternative nanoparticle placement procedures. The finding that the refractive index fluctuations in reflection are much larger than the ones observed in transmission also shows that, for a fixed nanoparticle distribution, the refractive indices obtained in transmission and reflection can significantly deviate from each other. Outside the homogeneous regime, the effective refractive index hence not merely depends on the microstate, but the same microstate can even be characterized by widely different effective refractive indices depending on the measure that is used (e.g. transmission or reflection). This shows that the concept of an effective refractive index looses most of its physical meaning outside the homogeneous regime.

To also gain insight into the physical mechanisms that lead to the refractive index fluctuations, I now investigate the distributions' density fluctuations on the scale of the wavelength. To this end, I used a cube with a side length of $a_{\text{cube}} = \lambda_0$ and systematically scanned this cube through the particle distributions. For each position (r_j) within the distributions, I then quantified the local volume fraction within the cube $(f_{\text{loc}}(r_j))$ by evaluating the partial volumes of all nanoparticles within the cube. These partial volumes are visualized by the blue areas of the nanoparticles in Fig. 3.2b). From this data, I subsequently quantified the distributions' density fluctuations as the standard deviation over all local volume fractions $(\Delta f_{\text{loc}} = \sigma(\{f_{\text{loc}}(r_j)\})$. Using this procedure, I obtained the results depicted in Fig. 3.4e. This figure shows that, for radii between $r_{\text{scat}} = 10$ nm and $r_{\text{scat}} = 80$ nm, the local density fluctuations increase close to linearly with r_{scat}. This suggests that the local density fluctuations are indeed one of the main drivers behind the refractive index fluctuations. Furthermore, in Sect. 3.3.3, I demonstrate that density fluctuations most likely also lead to incoherent scattering. This indicates that there is also an intrinsic connection between refractive index fluctuations and incoherent scattering.

Finally, the finding that the refractive index fluctuations increase drastically with the nanoparticles' size also has direct implications for the design of bulk optical nanocomposites. This is because the refractive indices of modern optical materials are accurate and reproducible beyond the 4th digit [4]. Nanocomposites are consequently only suitable replacements for conventional optical materials if their refractive index fluctuations are suppressed beyond this level. To quantify what particle size is required to fulfill this condition, I now additionally investigate the range of nanoparticle radii from $r_{\text{scat}} = 4$ nm up to $r_{\text{scat}} = 15$ nm. For this series of simulations, I fixed the distributions' widths and lengths at the values at which convergence was achieved for small particles ($w_{\text{dist}} = 3\omega_{\text{beam}}$ and $l_{\text{dist}} = 20 r_{\text{scat}}$; see Fig. 3.3). The data obtained from this series of simulations are shown in Fig. 3.4f. This figure demonstrates that, even in the range below $r_{\text{scat}} = 10$ nm, the fluctuations in the effective refractive index increase significantly with increasing nanoparticle size. Specifically, only for a radius of $r_{\text{scat}} = 4$ nm, which corresponds to a particle-to-wavelength ratio of $\frac{d_{\text{scat}}}{\lambda_0} = \frac{1}{100}$ the fluctuations in real$(n_{\text{eff}}^{\text{trans}})$ are reduced below 10^{-5}. In fact, the red line in Fig. 3.4f shows that this is also the nanoparticle size at which the local density fluctuations become negligible. This demonstrates that the nanoparticle sizes in bulk optical nanocomposites should ideally be below $d_{\text{scat}} = \frac{\lambda_0}{100}$ to suppress the refractive index fluctuations and hence make nanocomposites suitable replacements for conventional optical materials. In fact, this also shows that the concept of an effective refractive index already starts to loose its physical validity around $d_{\text{scat}} = \frac{\lambda_0}{100}$.

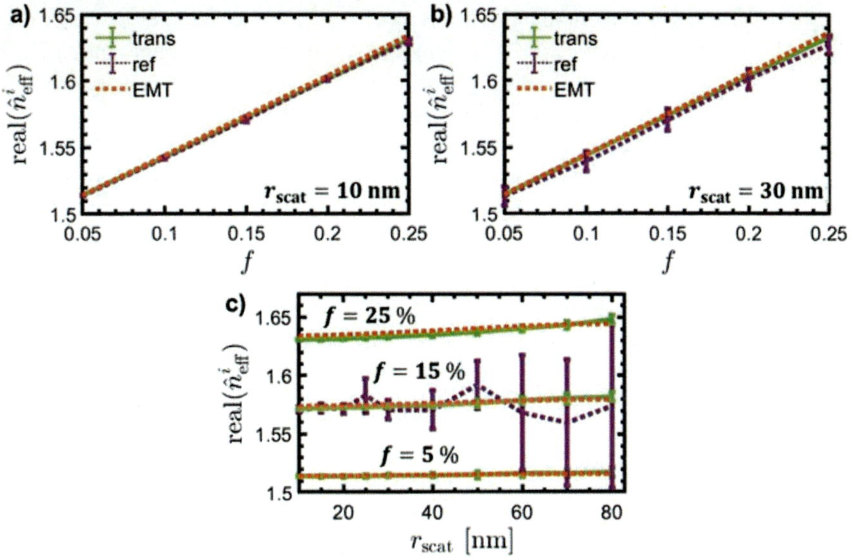

Fig. 3.5 Ensemble averages of the real part of the effective refractive index (real($\hat{n}^i_{\mathrm{eff}}$)) over the volume fraction (f) for nanoparticle radii of (**a**) $r_{\mathrm{scat}} = 10$ nm and (**b**) $r_{\mathrm{scat}} = 30$ nm. Both plots present the data obtained from the transmitted component, the reflected component, and the EMT ($i \in$ {trans, ref, EMT}). The ensemble averages denote the mean values of different random distributions (microstates) and the errorbars correspond to the standard deviations ($\Delta n^i_{\mathrm{eff}} = \sigma(\{n^i_{\mathrm{eff}}\})$. **c** Ensemble averages of the real part of the effective refractive index (real($\hat{n}^i_{\mathrm{eff}}$)) over the nanoparticle radius (r_{scat}) at volume fractions of $f = 5\,\%$, $f = 15\,\%$, and $f = 25\,\%$ ($i \in$ {trans, ref, EMT}). In reflection, only the results for $f = 15\,\%$ are presented to improve clarity. All remaining data are depicted in the appendix (Sect. A.1.5). Reproduced from [14]

3.3.2 *Ensemble Averages—The Real Part of the Effective Refractive Index*

The finding that, outside the homogeneous regime, different microstates can be characterized by drastically different effective refractive indices shows that it is necessary to distinguish a single microstate from the ensemble average, which can be obtained by averaging over a large number of different microstates. In fact, the effective medium theory (EMT) cannot account for statistical fluctuations (see Sect. 2.2). However, ideally it should be able to accurately predict the ensemble averages. In this section, I investigate whether this is the case for the real part of the effective refractive index. This also allows me to evaluate whether the EMT is a reliable tool for the design of nanocomposites with tailored refractive indices. I analyze the imaginary part in Sect. 3.3.3. Note that, from now on, I include error bars in all figures to visualize the fluctuations in the effective refractive index ($\Delta n^i_{\mathrm{eff}}$).

As the first step of my analysis of the real part, Fig. 3.5a shows that the ensemble averages of real($\hat{n}_{\mathrm{eff}}$) obtained in both reflection and transmission coincide at a radius of $r_{scat} = 10\,\text{nm}$. It is evident that this holds true for all volume fractions between $f = 5\,\%$ and $f = 25\,\%$. However, Fig. 3.5b demonstrates that this is no longer the case at a larger radius of $r_{scat} = 30\,\text{nm}$. This highlights that, for larger radii, even the ensemble averages depend on the measure that is used (e.g reflection and transmission). In this regime, there is consequently no effective refractive index that fully quantifies the material's optical response. It is rather necessary to use different values to describe different properties even if the ensemble averages are considered. Finally, Fig. 3.5a, b show that, in transmission, the EMT accurately predicts the ensemble averages for the real part at both $r_{scat} = 10\,\text{nm}$ and $r_{scat} = 30\,\text{nm}$. Only for the larger volume fractions, the EMT slightly overestimates real$(\hat{n}_{\text{eff}}^{\text{trans}})$. However, even for the largest volume fraction of $f = 25\,\%$, it remains accurate within $0.2\,\%$. In the appendix (Sect. A.1.3), I discuss that the slight tendency of the EMT to overestimate the real part of the effective refractive index is possibly caused by the fact that the distributions' interfaces are no longer well-defined for larger nanoparticles. In doing so, I show that a perfect match between the EMT and the numerical results can be achieved by introducing an effective interface.

As the second step, to elucidate the influence of the nanoparticle radius in detail, I selected three volume fractions ($f = 5\,\%$, $f = 15\,\%$, and $f = 25\,\%$) and systematically varied r_{scat} between 10 nm and 80 nm. The results of this analysis are shown in Fig. 3.5c. This figure confirms that, in transmission, the EMT generally predicts the real part of the effective refractive index with good accuracy. It is evident that this is true for all radii. As aforementioned, only at the maximum volume fraction of $f = 25\,\%$, the EMT's accuracy is slightly impaired. Furthermore, in the appendix (Sect. A.1.4), I show that the effective refractive indices obtained from the equation for metamaterials agree perfectly with the values retrieved from the transmitted component. This remains true even for the largest radius of $r_{scat} = 80\,\text{nm}$ and hence shows that the nanoparticles' magnetic dipole response is negligible for all radii included in this analysis. The finding that real$(\hat{n}_{\text{eff}}^{\text{trans}})$ begins to depend on the radii for larger values of r_{scat} (Fig. 3.5c) can consequently be attributed to the influence of the electric dipole resonance. However, the real part of the effective refractive index is essentially independent of r_{scat} for $r_{scat} < 20\,\text{nm}$. In this regime, the so-called Mie resonances [19] can consequently not be used to tailor the effective refractive index.

Figure 3.5c also demonstrates that large refractive index fluctuations are again observed in reflection. This can be directly seen from the purple line in Fig. 3.5c, which shows that the corresponding error bars span over a range of more than 0.1 refractive index units for the largest radii. Furthermore, in this regime, even the ensemble averages fluctuate strongly around the values predicted by the EMT. In the appendix (Sect. A.1.5), I show that strong fluctuations are also observed for all other volume fractions. These strong refractive index fluctuations demonstrate that, in reflection, the concept of an effective refractive index loses its applicability for radii over $r_{scat} = 30\,\text{nm}$. In this regime, the concept of an effective refractive index is consequently only useful for describing the transmitted component. However, even for this component, the prediction obtained from the EMT must be treated as an average measure that is obtained by averaging over a sufficiently large number of microstates, or alternatively, a sufficiently large volume.

In summary, the results presented in this section show that, if the values obtained from the EMT are treated as ensemble averages, the EMT generally predicts the real part of the effective refractive index with high accuracy (Fig. 3.5). In fact, I have shown that this remains true even at high volume fractions, for which multiple scattering plays a major role. Finally, nanocomposites that can be used as replacements for conventional optical materials must be located in the regime in which the refractive index fluctuations become negligible ($\frac{d_{\mathrm{scat}}}{\lambda_0} \leq \frac{1}{100}$; see Sect. 3.3.1). In this regime, the effective refractive indices of all microstates hence essentially coincide with the ensemble average. Therefore, the results presented in this section show that the EMT is an accurate tool for the design of optical nanocomposites with specifically tailored refractive indices. However, my results also demonstrate that the so-called Mie resonances cannot be used to tailor the effective refractive indices of bulk optical nanocomposites. This is because nanoparticle sizes of $d_{\mathrm{scat}} \leq \frac{\lambda_0}{100}$ are well below the threshold at which the Mie resonances begin to affect the effective refractive indices

3.3.3 The Imaginary Part of the Effective Refractive Index—The Influence of Incoherent Scattering

Besides the real part of the effective refractive index, its imaginary part is also a key quantity that must be considered for the design of optical nanocomposites. This is because, the imaginary part quantifies the overall amount of attenuation in the material. In fact, attenuation is a major issue for materials outside the homogeneous regime, because incoherent scattering [5, 6] plays a key role as a loss mechanism for such materials. I use the term incoherent scattering since, microscopically, all electromagnetic charges scatter electromagnetic waves. Therefore, it is necessary to distinguish the coherent part of the electromagnetic field, which is the propagating beam itself, from its incoherent part, which is formed by the radiation that is incoherently scattered out of the coherent part [5, 6, 33]. Incoherent scattering hence attenuates a beam that propagates through a material by redistributing a part of the beam's power flux in other directions. Incoherent scattering consequently leads to an increase of the imaginary part of the effective refractive index even if non-absorbing materials are used [5]. In imaging systems, incoherent scattering directly leads to a loss of contrast and hence severely reduces image quality.

As the first step in my analysis of the imaginary part, I again fixed the nanoparticle radii at $r_{\mathrm{scat}} = 10\,\mathrm{nm}$ and $r_{\mathrm{scat}} = 30\,\mathrm{nm}$ and investigate the influence of the volume fraction. Accordingly, Fig. 3.6a, b present the ensemble averages of $(\mathrm{imag}(n_{\mathrm{eff}}^{\mathrm{ref}}))$ for both radii over the volume fraction. These figures demonstrate that, in transmission, the imaginary part depends only weakly on the volume fraction. Moreover, the EMT provides an accurate prediction at small volume fractions, but overestimates the imaginary part at larger volume fractions for the transmitted component. In contrast to this behavior, the imaginary part obtained in reflection increases quickly with the volume fraction. For this component, the imaginary part generally also fluctuates

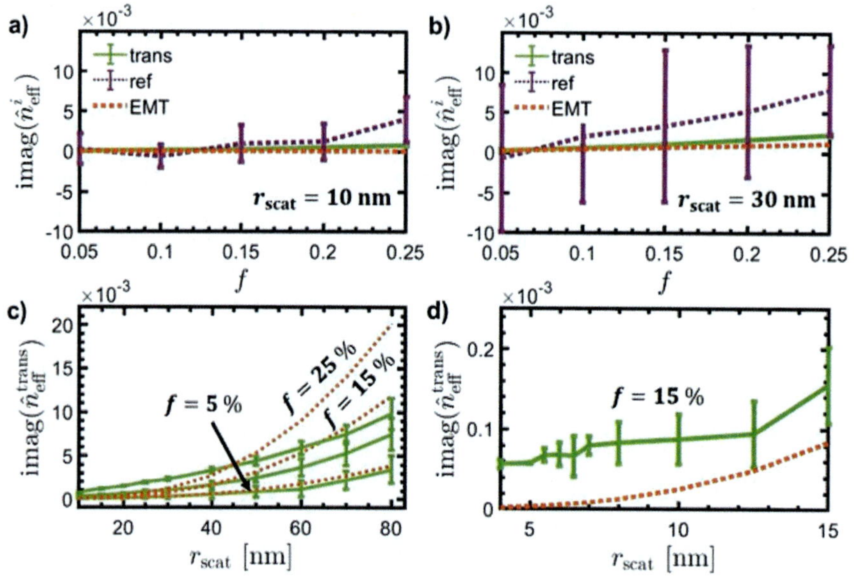

Fig. 3.6 Ensemble average of the imaginary part of the effective refractive index (imag($\hat{n}^i_{\text{eff}}$)) over the volume fraction (f) for two nanoparticle radii of (**a**) $r_{\text{scat}} = 10\,\text{nm}$ and (**b**) $r_{\text{scat}} = 30\,\text{nm}$. Both plots present the data obtained from the transmitted component, the reflected component, and the EMT ($i \in \{\text{trans, ref, EMT}\}$). The ensemble averages denote the mean values of different random distributions (microstates) and the errorbars correspond to the standard deviations ($\Delta n^i_{\text{eff}} = \sigma(\{n^i_{\text{eff}}\})$). **c** Ensemble average of the imaginary part of the effective refractive index in transmission (imag($\hat{n}^{\text{trans}}_{\text{eff}}$)) over the nanoparticle radius (r_{scat}) at volume fractions of $f = 5\,\%$, $f = 15\,\%$, and $f = 25\,\%$ ($i \in \{\text{trans, ref, EMT}\}$). All data obtained in reflection is presented in the appendix (Sect. A.1). **d** imag($\hat{n}^{\text{trans}}_{\text{eff}}$) over r_{scat} at a fixed volume fraction of $f = 15\,\%$ for nanoparticle radii between $4\,\text{nm} \leq r_{\text{scat}} \leq 15\,\text{nm}$. Reproduced from [14]

heavily and does not agree well with the values obtained from the EMT. In fact, it can be seen from both Fig. 3.6a, b that the imaginary part in reflection (imag($\hat{n}^{\text{ref}}_{\text{eff}}$)) even takes negative values for both $r_{\text{scat}} = 10\,\text{nm}$ and $r_{\text{scat}} = 30\,\text{nm}$. This can be attributed to plane waves with oblique incidence angles that are scattered into the normal direction.

As the second step, to again systematically analyze the influence of the nanoparticle radius, Fig. 3.6c presents imag($\hat{n}^{\text{trans}}_{\text{eff}}$) over r_{scat} for volume fractions of $f = 5\,\%$, $f = 15\,\%$, and $f = 25\,\%$. All data obtained in reflection are included in the appendix (Sect. A.1.5) because they fluctuate heavily. Figure 3.6c shows that the EMT predicts the imaginary part of the transmitted component with decent accuracy at low volume fractions and radii. However, upon increasing either of these quantities, the EMT looses its accuracy. This demonstrates that statistical differences in the nanoparticle distributions play a key role in these regimes. This follows from the fact that the EMT can only account for the influence of the particle size on the dipole polarizability (see (2.11)) but not for statistical deviations from perfect random distributions (see

Sect. 2.2). The behavior of the imaginary part observed in Fig. 3.6c consequently also shows that both the particle size itself and statistical influences, critically determine the total amount of incoherent scattering. In fact, this also demonstrates that both these factors directly cause the breakdown of the concept of an effective refractive index.

In general, the onset of incoherent scattering adds another level on which the concept of an effective refractive index breaks down. This is because incoherent scattering gradually distorts a beam of light (see Fig. 3.1) and the imaginary part can therefore no longer be used to determine the absorption rate. In fact, it also provides no information about the appearance of the material. A material with a high imaginary part of the effective refractive index can either appear completely black or opaque. In the former case, absorption is the dominant loss mechanism, whereas incoherent scattering dominates over absorption in the latter case. Moreover, I demonstrate in the appendix (Sect. A.1.6) that incoherent scattering also causes the effective refractive indices obtained in both reflection and transmission to depend on the distributions' widths (w_{dist}), even if they are a factor of 10 wider than the beam (ω_{beam}). This can be also be attributed to incoherent scattering, because light that is scattered out of the beam can lead to the emergence of long-ranged modes that propagate along the full width of the distribution. This width dependence adds another major level on which the concept of an effective refractive index breaks down.

Finally, incoherent scattering is an immense issue for optical systems. This is because incoherent scattering leads to stray light and consequently a loss of contrast in imaging applications. For the design of bulk optical nanocomposites, it is therefore essential to keep incoherent scattering at negligible level. Since I have already shown that the amount of incoherent scattering is critically determined by the nanoparticle size, it is straightforward that minimizing the amount of incoherent scattering can only be achieved by using sufficiently small nanoparticles. To determine what nanoparticle sizes (d_{scat}) are required to fulfill this requirement, I now again use my data for small radii from Sect. 3.3.1. Accordingly, Fig. 3.6d presents imag($\hat{n}_{\text{eff}}^{\text{trans}}$) for radii from $r_{\text{scat}} = 4\,\text{nm}$ up to $r_{\text{scat}} = 15\,\text{nm}$ at a volume fraction of $f = 15\,\%$. This figure demonstrates that, even for radii below $r_{\text{scat}} = 10\,\text{nm}$, the amount of incoherent scattering increases with increasing radii. In fact, as predicted by the EMT, the imaginary part appears to converge to its minimum around a radius of $r_{\text{scat}} = 4\,\text{nm}$. It appears likely that this minimum corresponds to the resolution limit of my numerical procedure, which likely also explains the systematic shift between the EMT and the numerical data. Specifically, it appears likely that the plane wave at normal incidence always experiences minor losses because of the combination of a relatively tightly focused beam with a finite-sized structure.

The finding that a radius of $r_{\text{scat}} = 4\,\text{nm}$ is required to minimize the attenuation demonstrates that a size-to-wavelength ratio around or ideally below $\frac{d_{\text{scat}}}{\lambda_0} = \frac{1}{100}$ is required for bulk optical nanocomposites. As discussed in Sect. 3.3.1, this is also the size at which refractive index fluctuations become negligible. Therefore, I conclude that nanocomposites can indeed replace conventional optical materials, but only for nanoparticle sizes below $d_{\text{scat}} = \frac{\lambda_0}{100}$. Furthermore, it is essential that the nanoparticles are well-incorporated into the host matrix, since defects in the polymer matrix or nanoparticle agglomerates act as defects that scatter light themselves. As I discussed in Sect. 2.3, this is a highly challenging task, which requires a careful optimization of the materials' chemical composition and the synthesis process. Finally, since nanoparticles with a size of $d_{scat} = \frac{\lambda_0}{100}$ are much smaller than the nanoparticles that have been used in most experimental works, this also explains why nanocomposites have mostly only been considered for thin coatings until now (see Chap. 1).

3.4 Effective Medium Regimes

My analysis of the transition from homogeneous to heterogeneous materials in the previous sections has shown that the properties of an optical material depend heavily on the size of the distinct building blocks that make up the material. Therefore, depending on the size of the building blocks, different effective medium regimes must be distinguished. In the following, I discuss these regimes and their practical implications in detail. This section combines the findings from the three dimensional numerical simulations discussed in the previous section with an analysis that was published in Optical Materials Express [34].

The different effective medium regimes are summarized in Fig. 3.7. First, for materials in the homogeneous regime, the (effective) refractive index fully quantifies the optical response, in that, the effective refractive index is independent of the measure that is used (e.g. reflection or transmission). Furthermore, in this regime, both the fluctuations in the effective refractive index and the amount of incoherent scattering remain negligible. In addition, the materials' interfaces are well-defined on the scale of the wavelength (λ). This is the regime in which nanocomposites can be used as bulk optical materials. As I showed in the previous section, this homogeneous regime is generally only reached for scatterer sizes ($d_{scat} = 2r_{scat}$) below $\frac{\lambda}{100}$. Upon increasing the scatterer sizes, the material then transitions out of the homogeneous regime and into the restricted effective medium regime. This is the regime in which the concept of an effective refractive index gradually breaks down on different levels. The yellow panel in Fig. 3.7 gives a detailed overview over all these levels as a summary of my findings from the previous sections. This overview highlights that the concept of an effective refractive index can describe some properties of a material in the restricted effective medium regime, but also that the applicability of any effective refractive index is severely restricted compared to the homogeneous regime. Finally, materials composed of very large scatterers are located in the heterogeneous regime. Materials in this regime are highly opaque because incoherent scattering causes a beam of light to quickly lose its coherent and directional character. The effective refractive index has hence (Fig. 3.8) completely lost its physical meaning. I directly visualize this relationship below. The strong increase of the amount of incoherent scattering that is observed for the largest particle radii in Fig. 3.6 indicates that a scatterer size of $d_{scat} = \frac{\lambda}{5}$ appears to be a reasonable threshold for the boundary between the restricted effective medium and the heterogeneous regimes. However, I emphasize that there is always a gradual transition between all regimes. In fact,

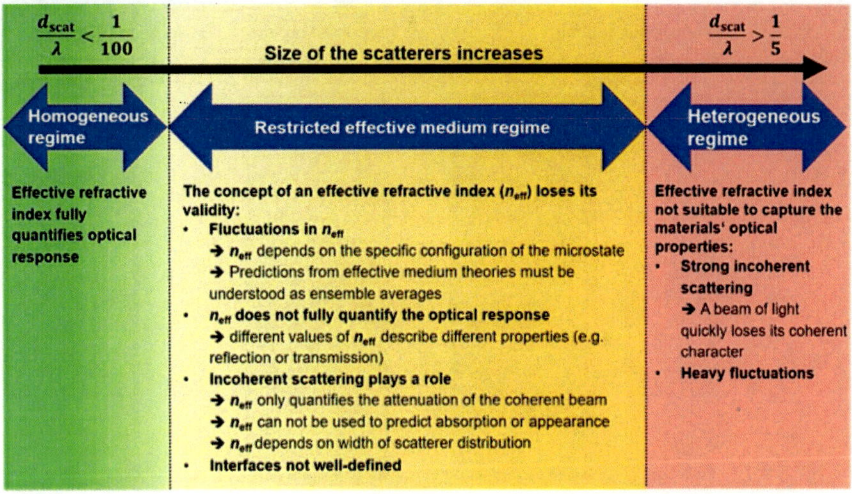

Fig. 3.7 Effective medium regimes: For materials in the homogeneous regime the (effective) refractive index fully quantifies the material's response. In contrast, in the restricted effective medium regime the concept of an effective refractive index breaks down. The color gradients indicate that there are no sharp boundaries between the regimes, but rather a gradual transition

since incoherent scattering is a process that gradually attenuates a beam of light, the thickness of the materials is also a key quantity that must be considered in practice. This is also why nanocomposites that include nanoparticles larger than $d_{\text{scat}} = \frac{\lambda}{100}$ can be used as thin transparent coatings, but not bulk optical materials.

Finally, to directly visualize the differences between the different effective medium regimes, I performed 2D finite element method (FEM) simulations using the commercially available software JCMsuite. For this purpose, I chose 2D simulations, because they allow me to model much larger propagation distances than the 3D approach from the previous section. For all simulations presented here, I fixed the volume (area) fraction at $f = 40\,\%$ and used a wavelength of $\lambda_0 = 0.8\,\mu\text{m}$. Furthermore, as a benchmark, I first simulated the propagation of a beam of light through a homogeneous material with a fixed refractive index. The spatial intensity distribution obtained from this simulation is presented in the left panel of Fig. 3.8. Second, to visualize the propagation of the beam through a nanocomposite in the homogeneous regime, I performed a simulation for a nanoparticle size of $d_{\text{scat}} = 6\,\text{nm}$. This value corresponds to a size-to-wavelength ratio $\frac{d_{\text{scat}}}{\lambda_0} = 0.0075$ and is hence well below the threshold of $\frac{d_{\text{scat}}}{\lambda_0} = \frac{1}{100}$ at which the material transitions into the restricted effective medium regime. The corresponding panel in Fig. 3.8 shows that, for this particle size, the beam of light indeed remains completely intact. This visualizes that incoherent scattering remains negligible and hence illustrates that nanocomposites in this regime can be used as homogeneous optical materials. In contrast, for a particle size of $d_{\text{scat}} = 100\,\text{nm}$, which corresponds to a size-to-wavelength ratio of $\frac{d_{\text{scat}}}{\lambda_0} = \frac{1}{8}$, incoherent scattering quickly separates the beam into different bundles (yellow panel

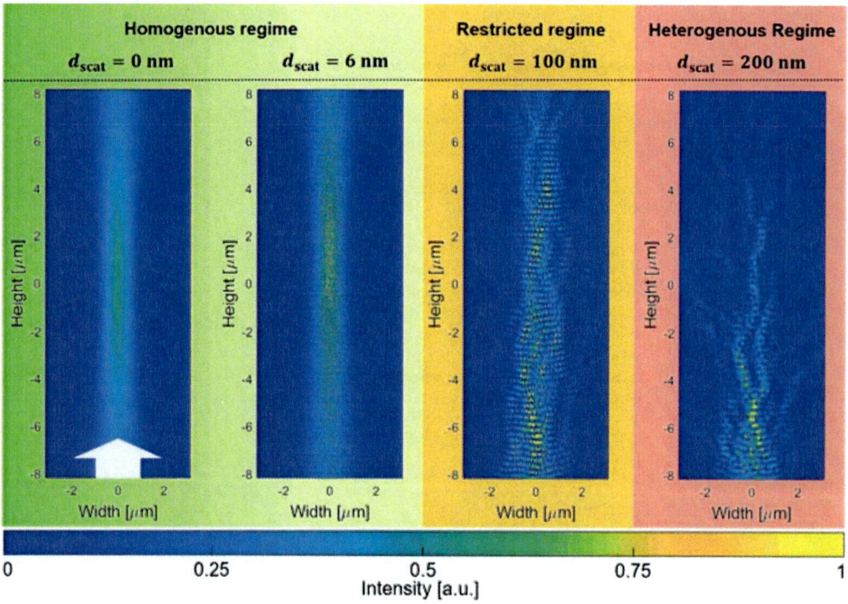

Fig. 3.8 Spatial intensity distribution from 2D FEM simulations to visualize the propagation of a focused beam of light ($\lambda_0 = 800$ nm) in the different effective medium regimes. TiO_2 was used as the nanoparticle material and PMMA as the host matrix. The volume (area) fraction remained fixed at 40 %. The colorbar for all images is shown at the bottom of the figure

in Fig. 3.8). However, it is evident that the beam still maintains its directionality to a certain degree. This illustrates that, in this regime, incoherent scattering gradually attenuates a beam of light. The concept of an effective refractive index can hence still be useful for describing the transmission of the coherent part of the beam though the material. Finally, for a particle size of $d_{scat} = 200$ nm, which corresponds to a size-to-wavelength ratio of $\frac{d_{scat}}{\lambda_0} = \frac{1}{4}$, the directionality is lost within a few micrometers. This can be directly seen from the red panel in Fig. 3.8. In this regime, a beam of light consequently quickly looses its coherent character and the concept of an effective refractive index has fully lost its applicability.

Nanocomposites are only suitable for optical systems in the homogeneous regime

In conclusion, in this chapter, I have shown that for nanocomposites to become suitable for bulk optical applications, the nanoparticles' size must not exceed a hundredth of the wavelength. To cover the entire visible spectral range with $0.4\,\mu m \leq \lambda \leq 0.8\,\mu m$, the nanoparticles' sizes must hence be below 4 nm to avoid refractive index fluctuations and incoherent scattering at all wavelengths. Since most experimental works and also the commercially available materials [35–37] relied on larger nanoparticles, this explains why these state-of-the-art materials are only used for thin coatings. Furthermore, I have shown using full wave optical simulations that the Maxwell-Garnett-Mie EMT (see (2.10) and (2.11)) is a highly accurate tool for predicting the real part of the effective refractive index. In the following chapters, I therefore use this EMT to investigate the potential of nanocomposites in the homogeneous regime as next-generation optical materials.

References

1. D. Aspnes, Local-field effects and effective-medium theory: a microscopic perspective. Am. J. Phys. **50**(8), 704–709 (1982)
2. H. Fearn, D.F. James, P.W. Milonni, Microscopic approach to reflection, transmission, and the Ewald-Oseen extinction theorem. Am. J. Phys. **64**(8), 986–995 (1996)
3. V.C. Ballenegger, T. Weber, The Ewald-Oseen extinction theorem and extinction lengths. Am. J. Phys. **67**(7), 599–605 (1999)
4. P. Hartmann, R. Jedamzik, S. Reichel, B. Schreder, Optical glass and glass ceramic historical aspects and recent developments: a Schott view. Appl.Opt. **49**(16), D157–D176 (2010)
5. C.F. Bohren, Applicability of effective-medium theories to problems of scattering and absorption by nonhomogeneous atmospheric particles. J. Atmosp. Sci. **43**(5), 468–475 (1986)
6. L.L. Foldy, The multiple scattering of waves. i. general theory of isotropic scattering by randomly distributed scatterers. Phys. Rev. **67**(3–4), 107–119 (1945)
7. M.I. Mishchenko, L. Liu, D.W. Mackowski, B. Cairns, G. Videen, Multiple scattering by random particulate media: exact 3D results. Opt. Exp. **15**(6), 2822–2836 (2007)
8. M.I. Mishchenko, L.D. Travis, T-matrix computations of light scattering by large spheroidal particles. Opt. Commun. **109**(1), 16–21 (1994)
9. A. Macke, M.I. Mishchenko, K. Muinonen, B.E. Carlson, Scattering of light by large nonspherical particles: ray-tracing approximation versus T-matrix method. Opt. Lett. **20**(19), 1934–1936 (1995)
10. Z. Cui, Y. Han, Q. Xu, Numerical simulation of multiple scattering by random discrete particles illuminated by Gaussian beams. J. Opt. Soc. Am. A **28**(11), 2200–2208 (2011)
11. L. Pattelli, A. Egel, U. Lemmer, D.S. Wiersma, Role of packing density and spatial correlations in strongly scattering 3D systems. Optica **5**(9), 1037–1045 (2018)
12. I. Fernandez-Corbaton, D. Beutel, C. Rockstuhl, A. Pausch, W. Klopper, Computation of electromagnetic properties of molecular ensembles. Chem. Phys. Chem. **21**, 878 (2020)
13. A. Egel, L. Pattelli, G. Mazzamuto, D.S. Wiersma, U. Lemmer, CELES: CUDA-accelerated simulation of electromagnetic scattering by large ensembles of spheres. J. Quant. Spect. Radiative Transfer **199**, 103–110 (2017)
14. D. Werdehausen, X.G. Santiago, S. Burger, I. Staude, T. Pertsch, C. Rockstuhl, M. Decker, Modelling optical materials at the single scatterer level—The transition from homogeneous to heterogeneous materials. Adv. Theory Simul. **3**, 2000192 (2020)
15. X.G. Santiago, M. Hammerschmidt, S. Burger, C. Rockstuhl, I. Fernandez-Corbaton, L. Zschiedrich, Decomposition of scattered electromagnetic fields into vector spherical wave functions on surfaces with general shapes. Phys. Rev. B **99**(4), 045406 (2019)
16. B. Freireich, M. Kodam, C. Wassgren, An exact method for determining local solid fractions in discrete element method simulations. AIChE J. **56**(12), 3036–3048 (2010)
17. F. Chen, P.R. Bakic, A.D.A. Maidment, S.T. Jensen, X. Shi, D.D. Pokrajac, Description and characterization of a novel method for partial volume simulation in software breast phantoms. IEEE Trans. Med. Imaging **34**(10), 2146–2161 (2015)
18. A. Rahbari, R. Hens, I.K. Nikolaidis, A. Poursaeidesfahani, M. Ramdin, I.G. Economou, O.A. Moultos, D. Dubbeldam, T.J.H. Vlugt, Computation of partial molar properties using continuous fractional component Monte Carlo. Molecular Phys. **116**(21–22), 3331–3344 (2018)
19. M. Decker, I. Staude, Resonant dielectric nanostructures: a low-loss platform for functional nanophotonics. J. Opt. **18**(10), 103001 (2016)

20. N. Sultanova, S. Kasarova, I. Nikolov, Dispersion properties of optical polymers. Acta Physica Polonica-Ser. A General Phys. **116**(4), 585 (2009)
21. S. Li, M. Meng Lin, M.S. Toprak, D.K. Kim, M. Muhammed, Nanocomposites of polymer and inorganic nanoparticles for optical and magnetic applications. Nano Rev. **1**(1), 5214 (2010)
22. P. Tao, Y. Li, A. Rungta, A. Viswanath, J. Gao, B.C. Benicewicz, R.W. Siegel, L.S. Schadler, TiO2 nanocomposites with high refractive index and transparency. J. Mater. Chem. **21**(46), 18623–18629 (2011)
23. J.L.H. Chau, Y.-M. Lin, A.-K. Li, W.-F. Su, K.-S. Chang, S.L.-C. Hsu, T.-L. Li, Transparent high refractive index nanocomposite thin films. Mater. Lett. **61**(14–15), 2908–2910 (2007)
24. S. Lee, H.-J. Shin, S.-M. Yoon, D.K. Yi, J.-Y. Choi, U. Paik, Refractive index engineering of transparent ZrO2-polydimethylsiloxane nanocomposites. J. Mater. Chem. **18**(15), 1751–1755 (2008)
25. C. Lü, Z. Cui, Z. Li, B. Yang, J. Shen, High refractive index thin films of ZnS/polythiourethane nanocomposites. J. Mater. Chem. **13**(3), 526–530 (2003)
26. C. Lü, Z. Cui, Y. Wang, Z. Li, C. Guan, B. Yang, J. Shen, Preparation and characterization of ZnS-polymer nanocomposite films with high refractive index. J. Mater. Chem. **13**(9), 2189–2195 (2003)
27. C. Lü, C. Guan, Y. Liu, Y. Cheng, B. Yang, PbS/polymer nanocomposite optical materials with high refractive index. Chem. Mater. **17**(9), 2448–2454 (2005)
28. C. Lü, B. Yang, High refractive index organic-inorganic nanocomposites: design, synthesis and application. J. Mater. Chem. **19**(19), 2884–2901 (2009)
29. R.J. Nussbaumer, W.R. Caseri, P. Smith, T. Tervoort, Polymer-TiO2 nanocomposites: a route towards visually transparent broadband UV filters and high refractive index materials. Macromolecular Mater. Eng. **288**(1), 44–49 (2003)
30. D.R. Smith, S. Schultz, P. Markoš, C.M. Soukoulis, Determination of effective permittivity and permeability of metamaterials from reflection and transmission coefficients. Phys. Rev. B **65**(19) (2002)
31. C. Rockstuhl, T. Zentgraf, H. Guo, N. Liu, C. Etrich, I. Loa, K. Syassen, J. Kuhl, F. Lederer, H. Giessen, Resonances of split-ring resonator metamaterials in the near infrared. Appl. Phys. B **84**(1–2), 219–227 (2006)
32. E. Ospadov, J. Tao, V.N. Staroverov, J.P. Perdew, Visualizing atomic sizes and molecular shapes with the classical turning surface of the Kohn-Sham potential. Proc. Natl. Acad. Sci. **115**(50), E11578–E11585 (2018)
33. M. Lax, Multiple scattering of waves. Rev. Modern Phys. **23**(4), 287–310 (1951)
34. D. Werdehausen, I. Staude, S. Burger, J. Petschulat, T. Scharf, T. Pertsch, M. Decker, Design rules for customizable optical materials based on nanocomposites. Opt. Mater. Exp. **8**(11), 3456 (2018)
35. Z. Chen, Pixelligent zirconia nano-crystals for OLED applications, in *White Paper* (2014)
36. D. Russel, A. Stabell. Scaling-up pixelligent nanocrystal dispersions, in *White Paper* (2016)
37. Z. Chen, J. Wang, Pixelligent internal light extraction layer for OLED lighting, in *White Paper* (2014)

Chapter 4
Nanocomposites as Tunable Optical Materials

In the previous chapter, I have shown that nanocomposites can be used as bulk optical materials. However, this is only possible in the homogeneous regime, which, for applications in the visible spectral range, is reached for particle sizes below 4 nm. The main degrees of freedom that remain available for the design of optical nanocomposites are hence only the constituent materials (host and nanoparticles) and their respective volume fractions. Therefore, the key question is whether significant benefits over conventional materials can be achieved with these degrees of freedom. To answer this question, I, in this chapter, investigate what range of optical properties can be achieved with nanocomposites in the homogeneous regime. Since I have already shown that the Maxwell-Garnett-Mie effective medium theory (EMT) is an accurate tool for the design of nanocomposites in the homogeneous regime, I first use this EMT (see (2.10) and (2.11)) to investigate the general potential of optical nanocomposites for a wide range of different materials. Subsequently, I present experimental data for specific materials and optical components to confirm these general findings. The results presented in Sect. 4.1 have been published in Optica [1], whereas the data in Sect. 4.3 has been published in a joint paper with the group of Prof. Harald Gießen of the University of Stuttgart [2].

4.1 Dispersion-Engineered Nanocomposites

As discussed in Sect. 2.1, the wavelength dependence of the refractive index ($n(\lambda)$) is commonly quantified using the refractive index at the d-line (n_d), the Abbe number (ν_d), and the partial dispersion ($P_{g,F}$). Adjusting these quantities is equivalent to tailoring the magnitude and the dispersion of $n(\lambda)$ (see Fig. 2.1 in Sect. 2.1.1).

D. Werdehausen, *Nanocomposites as Next-Generation Optical Materials*, Springer Series in Materials Science 316, https://doi.org/10.1007/978-3-030-75684-0_4

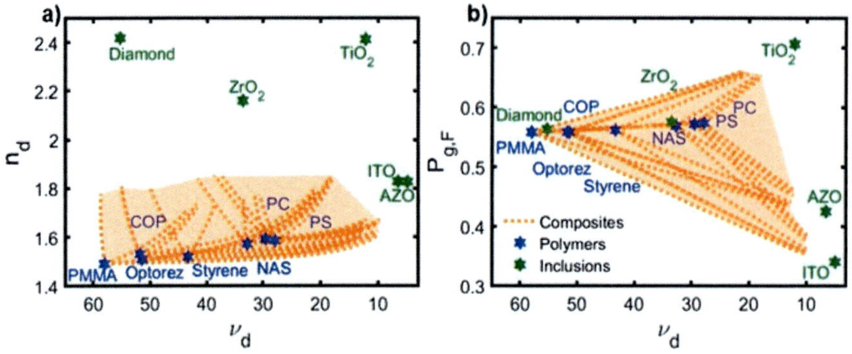

Fig. 4.1 **a** Refractive index at the d-line (n_d) and **b** partial dispersion ($P_{g,F}$) as a function of the Abbe number (ν_d) for widely used optical polymers (blue stars), different nanoparticle materials (green stars), and nanocomposites composed of the polymers and the nanoparticle materials for volume fractions of up to 35% (orange lines). The orange areas in both diagrams visualizes the range that is accessible by combining several nanoparticle materials in one host. (Acronyms: COP—Cyclic olefin copolymer, PMMA—Poly(methyl methacrylate), PC—Polycarbonate, PS—Polystyrene, ITO—Indium tin oxide,and AZO—Aluminum-doped zinc oxide). Reprinted with permission from [1] © The Optical Society

To visualize the range of n_d, ν_d, and $P_{g,F}$ values that is accessible with different kinds of optical materials, Fig. 4.1a, b depict the Abbe diagram (n_d as a function of ν_d) and the partial dispersion diagram ($P_{g,F}$ as a function of ν_d) for widely used optical polymers (blue stars) [3] and different nanoparticle materials (green stars). I focus on optical polymers because they today dominate many high-volume optical products, for example, smartphone cameras. This is because polymers are cheap, allow for efficient mass production, have a low weight, and high impact resistance [4, 5]. However, with respect to their optical properties, polymers also suffer from significant disadvantages. Namely, Fig. 4.1a shows that the magnitude of their refractive indices (n_d) is generally low even at low values of ν_d. Second, it can be seen from Fig. 4.1b that all polymers in this figure have similar partial dispersions ($P_{g,F}$). These constraints are limiting factors for optical systems relying solely on polymers [6]. This can be attributed to the fact that correcting the different monochromatic and chromatic aberrations, which affect the performance of optical systems, becomes increasingly difficult if the accessible range of n_d, ν_d, and $P_{g,F}$ values is restricted [7]. For example, materials with high refractive indices allow for correcting spherical aberration, wheras the combination of such materials with low-refractive-index materials is a powerful tool for reducing the Petzval field curvature [7]. The low refractive indices of polymers can hence be a limiting factor for the design of polymer-based systems. In fact, I discuss an intuitive example that highlights the impact high-refractive-index nanocomposites can have in optical design in Sect. 4.2. Finally, as I discuss in detail in Chap. 6, additional limitations arise for polymer-based systems when it comes to correcting chromatic aberrations to a very high degree. This is because polymers only cover a limited range in the partial dispersion diagram (see Fig. 4.1b).

To demonstrate that nanocomposites based on optical polymers allow for overcoming the disadvantages of polymers to a high degree, I first used the EMT to predict the properties of the full range of nanocomposites that are composed of the materials included in Fig. 4.1. Specifically, I investigated all possible combinations of the polymers as the host matrices and the nanoparticle materials as the inclusions. First, the orange lines in Fig. 4.1a, b demonstrate that incorporating one nanoparticle material into a polymer allows for tailoring the resulting nanocomposite's properties along a fixed line. It is evident that this line starts at the polymer itself and is directed towards the location of the nanoparticle material. Varying the nanoparticle material's volume fraction hence allows for tuning the nanocomposite's optical properties along this line. Note that I included volume fractions of up to $f = 35\%$ in Fig. 4.1, which is well within the range that has been achieved experimentally [8]. All orange lines in Fig. 4.1 together consequently show that adding one nanoparticle material into a polymer enlarges the accessible area of n_d, ν_d, and $P_{\mathrm{g,F}}$ values way beyond that of conventional polymers. In fact, it is possible to access widely different dispersion regimes by combining different materials. In addition, it is evident that nanocomposites extend to much higher refractive indices than conventional polymers.

The finding that tailoring the volume fractions allows for engineering their effective refractive indices along the orange lines in Fig. 4.1 shows that nanocomposites can serve as tailorable optical materials. Building on this finding, I can further generalize this approach by exploiting that, in principle, also multiple filler materials can be combined in the same host. The shaded orange areas in Fig. 4.1 show that this concept of combining multiple filler materials allows for continuously adjusting n_d, ν_d, and $P_{\mathrm{g,F}}$ within wide regions in both diagrams. I obtained these orange areas by first generalizing the Clausius-Mossotti equation and thus the Maxwell-Garnett-Mie EMT to two different kinds of inclusions (see [9] for the generalization of the Clausius-Mossotti equation). This allowed me to vary the different inclusions' volume fractions independently. In doing so, I maintained the constraint that the nanoparticles' overall volume fraction must not exceed $f = 35\%$. The large size of the orange areas in Fig. 4.1, which results from this procedure, shows that nanocomposites with multiple inclusions can provide an independent control over the magnitude and dispersion of the refractive index to a very large degree. Nanocomposites can consequently serve as dispersion-engineered materials that can be individually optimized within a wide region for each application at hand. These "dispersion engineering" capabilities of nanocomposites are one of the central findings of this book and the following sections and chapters evolve around how these capabilities can be used to enhance optical elements and systems. Accordingly, I define the term nanocomposite-enabled optical element or system as an optical device that exhibits a functionality or performance that is significantly enhanced compared to those that could be achieved with conventional materials.

4.2 Nanocomposite-Enabled Optical Elements and Systems

As aforementioned, the combination of materials that cover wide regions in the Abbe and partial dispersion diagrams is a powerful tool for correcting aberrations in optical systems [7]. For example, in Chap. 6, I demonstrate that the wide region covered by nanocomposites in the partial dispersion diagram in Fig. 4.1b makes them perfectly suited for correcting chromatic aberrations, that is, the design of achromatic optical systems. At this point, I instead focus on the finding that nanocomposites extend to much higher refractive indices than conventional polymers to briefly demonstrate the benefits of nanocomposites on one specific example. For this purpose, I used the commercial optical design software "OpticStudio" to design a lens that is made of a conventional polymer (PMMA) and a second lens, with the same specifications, that is made of a PMMA-based diamond nanocomposite at a volume fraction of $f = 35\%$. This nanocomposite corresponds to the endpoint of the orange line that starts at PMMA and is aimed at diamond in Fig. 4.1a. Comparing the two lenses' layouts in Fig. 4.2 shows that the nanocomposite-enabled lens is distinguished by an increased performance and a reduced thickness. The improved performance manifests itself in a significant reduction of the spot size (s) by more than 55%. This can be attributed to the higher refractive index of the nanocomposite, which reduces the lens' curvature and hence the amount of spherical aberration [7] compared to the conventional polymer.

The two lenses in Fig. 4.2 are, of course, just a first example for how nanocomposites could be used to enhance optical systems. In fact, the finding that the effective optical properties of nanocomposites can be continuously optimized within a wide

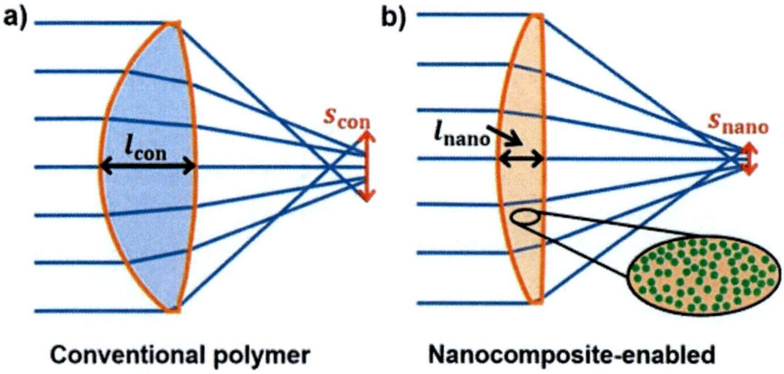

Fig. 4.2 Layout of **a** a lens made of conventional polymers and **b** a nanocomposite-enabled lens. Both lenses have a focal length of $f = 4$ mm and an aperture of $D = 4$ mm. The blue lines depict light rays that propagate through the lenses according to Snell's law. The comparison of both lenses shows that nanocomposite-enabled lens exhibits a higher performance $s_{nano} < s_{con}$ and is thinner $l_{nano} < l_{con}$ than the conventional lens. This can be attributed to the high refractive index of the nanocomposite. Both lenses were designed using the optical design software "OpticStudio", which relies on ray-tracing. Adapted with permission from [2] © The Optical Society

range shows that the use of nanocomposites in optical systems would provide optical designers with the possibility to use the materials' properties themselves as continuous degrees of freedom in the design process. The design of nanocomposite-enabled system would hence entail using the composition of nanocomposites, which is defined by their constituent materials and respective volume fractions, as powerful degrees of freedom that can be optimized to minimize the system's aberrations and hence maximize its performance. Note that, alternatively, such a performance benefit can generally also be leveraged to improve other specifications, for example, to reduce the system's size. Finally, the discussion about exploiting the degrees of freedom that become accessible with nanocomposites highlights that the EMT is an indispensable tool for the design of nanocomposite-enabled optical elements and systems. This is because the EMT provides a direct connection between the optimal optical properties (n_d, ν_d, and $P_\mathrm{g,F}$) and the actual composition of the nanocomposites. To this end, the EMT could, in the future, be directly implemented into commercial optical design solutions. In fact, building on the EMT, I develop more advanced concepts for how nanocomposites can enable a new generation of optical elements and systems in Chaps. 5 and 6.

4.3 Nanocomposites for 3D Printed Micro-optics

In this chapter, I have so far used the EMT to investigate the properties of a wide range of different nanocomposites. This has shown that nanocomposites can serve as dispersion-engineered materials. However, an experimental proof of principle for this approach is still missing. For this purpose, that is, to confirm my general findings and show that nanocomposites can indeed serve as tailorable optical materials, I now present experimental results for four different nanocomposites. To this end, I here focus on materials that can be used for 3D printing, specifically, femtosecond direct laser writing, which is a technology that has recently emerged as a powerful approach for 3D printing both optical [10–23] and other [24–31] devices on length scales from the submicrometer to the millimeter level. Briefly, this technology is based on scanning a tightly focused beam from a femtosecond laser through the volume of an initially liquid photoresist [32]. A photoresist is a material that generally contains a monomer and a photoinitiator, which activates the polymerization process when the photoresist is exposed to ultraviolet light. In femtosecond direct laser writing, the photoinitiator and the wavelength of the femtosecond laser must then be chosen such that the laser can active that photoinitiator only through two-photon absorption. Since two-photon absorption is a nonlinear process, which requires high intensities, it is locally restricted to a small voxel around the beam's focus. The photoresist is consequently only polymerized, that is, hardened within this voxel. Scanning the focused laser through the photoresist's volume then allows for fabricating 3D structures [32]. As such, femtosecond direct laser writing inherently is an additive manufacturing technology, which makes it ideally suited for rapid prototyping and providing a proof of principle for the concept of nanocomposite-enabled optical elements. In addition, femtosecond direct laser writing is also a powerful approach for fabricating novel micro-optical systems. This is because it provides the possibility to realize complex structures that can only be fabricated with much higher effort or cannot be realized at all with conventional substractive approaches [32, 33]. However, as a manufacturing technology that relies on polymer-based materials, such 3D printed optical elements and systems also suffer from the disadvantages of polymers, namely that they only cover a narrow range of optical properties. For these reasons, I here focus on materials that can be used as photoresists for femtosecond direct laser writing. In the following section, I then present experimental results for prototypical 3D printed nanocomposite-enabled optical elements. All results presented in this section were obtained in a close collaboration with the group of Prof. Harald Giessen at the University of Stuttgart and were recently published in a joint paper [2].

4.3.1 Nano-Inks for Femtosecond Direct Laser Writing

As the host material for our first three prototype nanocomposites ("nano-inks"), the photoresist "IP-Dip" (supplier: Nanoscribe GmbH)) was used. This material is a standard photoresist ("ink") for femtosecond direct laser writing [10, 15]. However, it has a low refractive index of merely $n_\mathrm{d} = 1.55$ at a Abbe number of $\nu_\mathrm{d} = 35.44$ [10, 15]. In fact, as discussed previously, hardened photoresists are polymers and hence also suffer from low refractive indices. To experimentally demonstrate that nanocomposites allow for overcoming this limitation and determine the accuracy of the EMT, commercially available ZrO_2 nanoparticles [34–36] were incorporated into IP-Dip at three different volume fractions between $f = 10\%$ and $f = 21\%$. As already discussed in Sect. 2.3, these commercially available ZrO_2 nanoparticles (supplier: Pixelligent Technologies LLC) are available with different capping layers, which are designed to achieve compatibility with a wide range of polymers. Therefore, a functionalization for which a good compatibility with IP-Dip was achieved could be selected (name of product: "PCPA"). Note that, with an average size of just below $d_\mathrm{scat} = 10\,\mathrm{nm}$, the commercial nanoparticles are still too large for bulk optical applications in the visible spectral range. However, since I have already shown in Sect. 3.3.2 that the real part of the effective refractive index is independent of the nanoparticle radius for $r_\mathrm{scat} < 20\,\mathrm{nm}$, these materials are perfectly suited for confirming the validity of the EMT. In addition, the impact of incoherent scattering in micro-optical components is reduced compared to conventional optical elements that operate on larger length scales, because the total amount of incoherent scattering increases with the propagation length. As I show below, nanocomposites based on these commercially available ZrO_2 nanoparticles can hence still be suitable for certain micro-optical applications.

For the synthesis of the prototypical nano-inks, the ZrO_2 nanoparticles, which were provided in a propylene glycol methyl ether acetate (PGMA) solution, were first mixed with the IP-Dip monomer. Second, the solvent was removed under reduced pressure. Third, for the initial systematic characterization, the final mixture was poly-

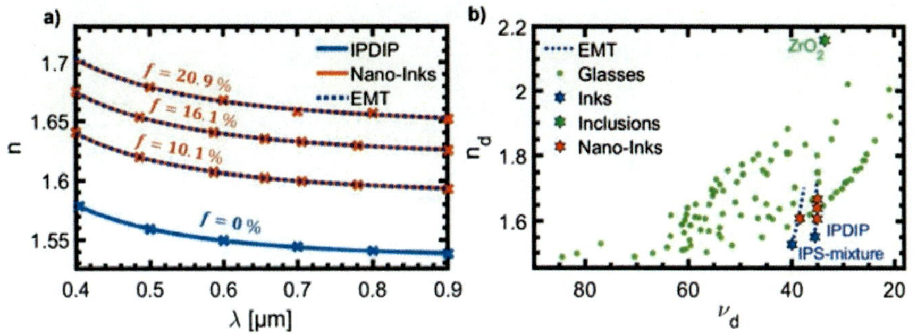

Fig. 4.3 **a** Refractive index as a function of the wavelength for IP-Dip (supplier: Nanoscribe GmbH)) as the host material (blue line) and three ZrO_2 nano-inks (nanoparticle supplier: Pixelligent Technologies LLC) based on IP-Dip (red lines). The crosses denote the measured data, whereas the red solid lines visualize the corresponding Cauchy fits (see (2.1)). The volume fractions (f) were determined from the MGM effective medium theory (EMT). **b** Abbe diagram including the IP-Dip based nano-inks and, additionally, a nano-ink that is based on a different host material (an IP-S [10] based polymer mixture). It is evident from both figures that there is an almost perfect match between the EMT and the experimental data

merized using UV light, that is, using conventional one photon absorption. Finally, the refractive indices of the resulting polymerized nano-inks were characterized using a commercial automated Pulfrich refractometer. Figure 4.3a presents the measured refractive indices of the nano-inks as a function of the wavelength. Note that I determined the volume fractions of the different nano-inks from the EMT by using f as a single free parameter. This fitting procedure, intuitively speaking, finds the location in the Abbe diagram that best matches the experimental data along the line in this diagram that is defined by the constituent materials (see Fig. 4.1a). I emphasize that using the volume fraction as a single free parameter cannot shift the trajectory of the line itself. This is because this line is fixed solely by the properties of the constituent materials.

Figure 4.3a presents the refractive indices of IP-Dip and our three nano-inks as a function of the wavelength. It is evident that these refractive index curves confirm the results from the previous sections. Namely, first, the experimental data show that increasing the nanoparticles' volume fraction systematically increases the refractive index. Second, it is evident that the refractive index profiles obtained from the EMT (see Sect. 2.2) almost perfectly match the experimental data. This confirms that the EMT is a highly accurate tool for the design of dispersion-engineered nanocomposites. As an additional visualization of all materials' properties, Fig. 4.3b depicts all nano-inks' locations in the Abbe diagram. This figure also includes a nano-ink that is based on the conventional photoresist "IP-S" (supplier: Nanoscribe GmbH) as the host material. For the synthesis of this nano-ink, the IP-S monomer was first mixed with 2-Hydroxy-3-phenoxypropylacrylat (HPPA) at a 1:1 ratio. This improved the blending of the nanoparticles into the final polymer matrix. Since this additional nano-ink is located at a higher Abbe number than all IP-Dip based nano-inks, this

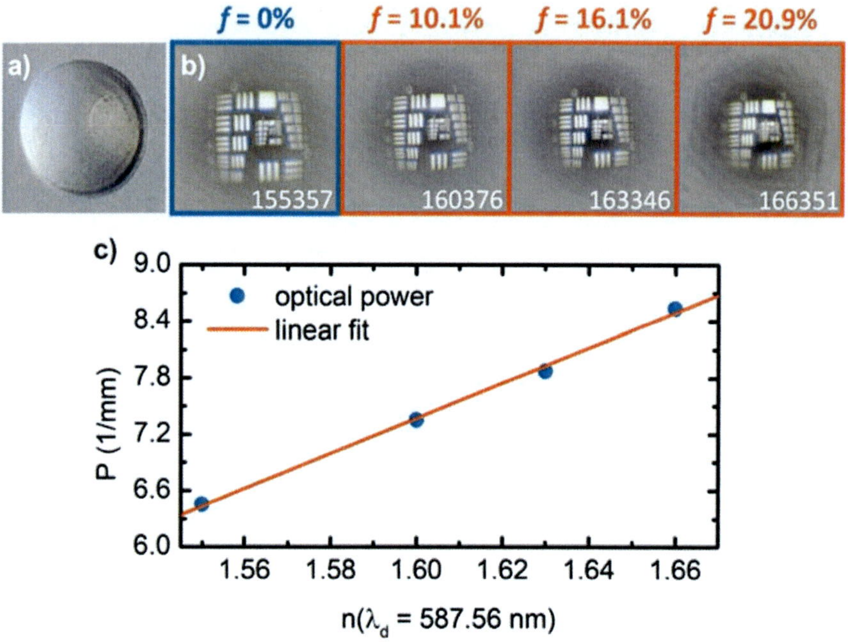

Fig. 4.4 **a** Microscope image of a lens made of the $f = 10.1\%$ IP-Dip based nano-ink. **b** Images of a USAF 1951 resolution test chart obtained through different lenses made of the IP-Dip based nano-inks (see Fig. 4.3). All lenses have a diameter of $A = 120$ μm and fixed radius of curvature of 100 μm. The numbers in the bottoms of the frames denote the materials' international glass code, which provides the position of each material in the Abbe diagram. The first three digits of this code correspond to the refractive index (n_d) and the final three digits denote the Abbe number (ν_d) [37]. **c** Measured refractive power ($P = \frac{1}{f}$) of all lenses as a function of their materials' refractive index at the d-line ($n_d = n(\lambda_d)$). Reprinted with permission from [2] © The Optical Society

additional nano-ink confirms that combining different materials allows for accessing different regions of the Abbe diagram. In addition, Fig. 4.3b also directly visualizes that EMT is able to predict the locations of all four nano-inks in the Abbe diagram with very high accuracy. Finally, Fig. 4.3b nicely illustrates that the nano-inks have much higher refractive indices than the two conventional photoresists. In fact, our nano-inks enter the region in the Abbe diagram that is normally only covered by optical glasses. The discussion in Sect. 4.2 illustrate that this could open up new design possibilities for 3D printed optical elements and systems.

The experimental results presented in this section provide a demonstration that it is indeed possible to develop nanocomposites into a material platform, which not only allows for accessing a wide range of optical properties but also enables the design of dispersion-engineered materials that can be specifically optimized for the application at hand. In the following section, I show that it is also possible to fabricate optical elements out of nanocomposites.

4.3.2 3D Printed Nanocomposite-Enabled Micro-optical Elements

The advantage of using photoresists, which were specifically designed for femtosecond direct laser writing, as the host material for our nano-inks is that our nano-inks can be directly used in this additive manufacturing process. Therefore, the rapid prototyping capabilities of femtosecond direct laser writing can be exploited to 3D print micro-optical components out of our nano-inks. This can provide an ideal first proof of principle that it is indeed possible to fabricate nanocomposite-enabled optical elements. Accordingly, I now present a systematic analysis for micro-optical lenses that are based on the three different IP-Dip based nano-inks. All devices presented in this section were fabricated and characterized by Ksenia Weber at the University of Stuttgart.

As a reference for the nanocomposite-lenses, Ksenia first used femtosecond direct laser writing to 3D print a single plano-convex lens with a diameter of 120 μm and a radius of curvature of 100 μm out of conventional IP-Dip. Subsequently, she printed three additional lenses, with the same dimensions, out of the different IP-Dip based nano-inks. To visualize the result of the printing process, an exemplary microscope image of the lens that was fabricated using the $f = 10.1\%$ nano-ink is depicted in Fig. 4.4a. All four lenses were fabricated using the commercial "Photonics Professional GT" 3D printer (supplier: Nanoscribe GmbH). This system uses a femtosecond laser, which operates at a wavelength of 780 nm and has a pulse duration in the order of 100 fs and achieves high speed scanning of the laser beam through the photoresists using two galvometric mirrors.

To systematically characterize the lenses, Ksenia first took images of an USAF 1951 resolution test chart through each of the four lenses. The object distance, that is, the distance between the lenses and the test chart was kept constant in all four cases. Focusing was achieved by changing the distance between the lens and the 20x (NA = 0.45) microscope objective, which, together with a CCD camera, was used to obtain the images. The corresponding images are presented in Fig. 4.4b. These images show that the size of the images on the sensor decreases with increasing volume fraction of the ZrO_2 nanoparticles. This is because the increase of the refractive index with increasing volume fraction increases the refractive power, that is, decreases the focal length. It is well known in optical design that, if the object distance remains unchanged, this leads to a decrease of the magnification [7]. Subsequently, Ksenia measured all lenses' refractive powers ($P = \frac{1}{f}$) to also quantify this relationship systematically. The results of these measurements are depicted in Fig. 4.4c. This figure shows that the refractive power indeed exhibits the expected linear behavior, which follows directly from the lensmaker's equation ($P \propto (n - 1)$).

The three lenses that Ksenia 3D printed out of our nano-inks provide an inital proof of principle for the concept of nanocomposite-enabled optical elements. However, I note again that the commercially available nanoparticles [34–36], which we used for the synthesis of the nano-inks, have sizes of around $d_{\text{scat}} = 10$ nm. Therefore, future work is required to go from these micro-optical lenses to much thicker macroscopic lenses. To this end, as I have shown in this chapter, sizes below $d_{scat} = 4\,\text{nm}$ are required to suppress incoherent scattering. Furthermore, as discussed in Sect. 2.3, it is critical to achieve a perfect blending of the nanoparticles into the polymer matrix. Fulfilling both these requirements to a very large degree will require significant future refinements of the materials and, simultaneously, also of the manufacturing processes. In fact, it is evident from Fig. 4.4b that incoherent scattering is also visible in the images that were recorded through the nanocomposite-lenses with the highest volume fractions. This illustrates that future refinements are indeed indispensable. Finally, the approach of combining nanocomposites with additive manufacturing holds an immense potential for the design of optical components as well as other devices. This is because, it allows for combining the immense design flexibility of additive manufacturing approaches [10–33] with the possibility to simultaneously tailor the materials themselves.

Nanocomposites unlock the material properties as continuous degrees of freedom

In conclusion, in this chapter, I have shown that nanocomposites in the homogeneous regime are a material platform that allows for precisely tailoring $n(\lambda)$ within a wide range. In fact, this range exceeds that of normal polymers by far. These materials consequently unlock the magnitude and dispersion of the effective refractive index as continuous degrees of freedom in the design of optical components. The exploitation of these degrees of freedom has subsequently lead me to the concept of nanocomposite-enabled optical elements. Finally, I have now demonstrated using both experimental results and large-scale numerical simulations that the EMT is a highly accurate tool for the design of dispersion-engineered nanocomposites. In the following sections and chapters, I therefore use the EMT to develop more advanced concepts for how nanocomposites can enable a new generation of optical systems.

References

1. D. Werdehausen, S. Burger, I. Staude, T. Pertsch, M. Decker, Dispersion-engineered nanocomposites enable achromatic diffractive optical elements. Optica **6**(8), 1031 (2019)
2. K. Weber, D. Werdehausen, P. Koenig, S. Thiele, M. Schmid, M. Decker, P.W. De Oliveira, A. Herkommer, H. Giessen, Tailored nanocomposites for 3D printed micro-optics. Opt. Mater. Exp. **10**(10), 2345 ((in press))
3. N. Sultanova, S. Kasarova, I. Nikolov, Dispersion properties of optical polymers. Acta Physica Polonica-Ser. A General Phys. **116**(4), 585 (2009)
4. N. Sultanova, S. Kasarova, I. Nikolov, Application of optical polymers in lens design, in *AIP Conference Proceedings*, vol. 1722.1 (2016), p. 230003
5. N. Sultanova, S. Kasarova, I. Nikolov, Advanced applications of optical polymers. Bulgarian J. Phys. **43**(3), 243–250 (2016)
6. P. Hartmann, R. Jedamzik, S. Reichel, B. Schreder, Optical glass and glass ceramic historical aspects and recent developments: a Schott view. Appl. Opt. **49**(16), D157–D176 (2010)
7. H. Gross, W. Singer, M. Totzeck, F. Blechinger, B. Achtner, *Handbook of Optical Systems*, vol. 1 (Wiley-VCH, Berlin, 2005)

8. P. Tao, Y. Li, A. Rungta, A. Viswanath, J. Gao, B.C. Benicewicz, R.W. Siegel, L.S. Schadler, TiO2 nanocomposites with high refractive index and transparency. J. Mater. Chem. **21**(46), 18623–18629 (2011)
9. V.A. Markel, Introduction to the Maxwell Garnett approximation: tutorial. J. Opt. Soc. Am. A **33**(7), 1244–1256 (2016)
10. T. Gissibl, S. Wagner, J. Sykora, M. Schmid, H. Giessen, Refractive index measurements of photo-resists for three-dimensional direct laser writing. Opt. Mater. Exp. **7**(7), 2293–2298 (2017)
11. T. Gissibl, S. Thiele, A. Herkommer, H. Giessen, Two-photon direct laser writing of ultracompact multi-lens objectives. Nat. Photon. **10**(8), 554 (2016)
12. S. Thiele, K. Arzenbacher, T. Gissibl, H. Giessen, A.M. Herkommer, 3D-printed eagle eye: Compound microlens system for foveated imaging. Sci. Adv. **3**(2), e1602655 (2017)
13. S. Thiele, C. Pruss, A.M. Herkommer, H. Giessen, 3D printed stacked diffractive microlenses. Opt. Exp. **27**(24), 35621 (2019)
14. M. Schmid, S. Thiele, A. Herkommer, H. Giessen, Three-dimensional direct laser written achromatic axicons and multi-component microlenses. Opt. Lett. **43**(23), 5837–5840 (2018)
15. M. Schmid, D. Ludescher, H. Giessen, Optical properties of photoresists for femtosecond 3D printing: refractive index, extinction, luminescence-dose dependence, aging, heat treatment and comparison between 1-photon and 2-photon exposure. Opt. Mater. Exp. **9**(12), 4564–4577 (2019)
16. K. Weber, F. Hütt, S. Thiele, T. Gissibl, A. Herkommer, H. Giessen, Single mode fiber based delivery of OAM light by 3D direct laser writing. Opt. Exp. **25**(17), 19672–19679 (2017)
17. A. Asadollahbaik, S. Thiele, K. Weber, A. Kumar, J. Drozella, F. Sterl, A.M. Herkommer, H. Giessen, J. Fick, Highly efficient dual-fiber optical trapping with 3D printed diffractive fresnel lenses. ACS Photon. **7**(1), 88–97 (2020)
18. T. Gissibl, S. Thiele, A. Herkommer, H. Giessen, Sub-micrometre accurate free-form optics by three-dimensional printing on single-mode fibres. Nat. Commun. **7**(1), 11763 (2016)
19. M.S. Rill, C. Plet, M. Thiel, I. Staude, G. von Freymann, S. Linden, M. Wegener, Photonic metamaterials by direct laser writing and silver chemical vapour deposition. Nat. Mater. **7**, 543 (2008)
20. M. Deubel, G. von Freymann, M. Wegener, S. Pereira, K. Busch, C.M. Soukoulis, Direct laser writing of three-dimensional photonic-crystal templates for telecommunications. Nat. Mater. **3**(7), 444–447 (2004)
21. M.F. Schumann, M. Langenhorst, M. Smeets, K. Ding, U.W. Paetzold, M. Wegener, All-angle invisibility cloaking of contact fingers on solar cells by refractive free-form surfaces. Adv. Opt. Mater. **5**(17), 1700164 (2020)
22. J. Fischer, M. Wegener, Three-dimensional optical laser lithography beyond the diffraction limit. Laser Photon. Rev. **7**(1), 22–44 (2020)
23. M.S. Rill, C. Plet, M. Thiel, I. Staude, G. von Freymann, S. Linden, M. Wegener, Photonic metamaterials by direct laser writing and silver chemical vapour deposition. Nat. Mater. **7**(7), 543–546 (2008)
24. M. Hippler, E.D. Lemma, S. Bertels, E. Blasco, C. Barner-Kowollik, M. Wegener, M. Bastmeyer, 3D scaffolds to study basic cell biology. Adv. Mater. **31**(26), 1808110 (2020)
25. T. Frenzel, M. Kadic, M. Wegener, Three-dimensional mechanical metamaterials with a twist. Sci. **358**(6366), 1072 (2017)
26. M. Kadic, T. Frenzel, M. Wegener, When size matters. Nat. Phys. **14**(1), 8–9 (2018)
27. M. Hippler, E. Blasco, J. Qu, M. Tanaka, C. Barner-Kowollik, M. Wegener, M. Bastmeyer, Controlling the shape of 3D microstructures by temperature and light. Nat. Commun. **10**(1), 232 (2019)
28. I. Fernandez-Corbaton, C. Rockstuhl, P. Ziemke, P. Gumbsch, A. Albiez, R. Schwaiger, T. Frenzel, M. Kadic, M. Wegener, New twists of 3D chiral metamaterials. Adv. Mater. **31**(26), 1807742 (2020)
29. M. Gernhardt, E. Blasco, M. Hippler, J. Blinco, M. Bastmeyer, M. Wegener, H. Frisch, C. Barner-Kowollik, Tailoring the mechanical properties of 3D microstructures using visible light post-manufacturing. Adv. Mater. **31**(30), 1901269 (2020)

30. L. Yang, A. Münchinger, M. Kadic, V. Hahn, F. Mayer, E. Blasco, C. Barner-Kowollik, M. Wegener, On the schwarzschild effect in 3D two-photon laser lithography. Adv. Opt. Mater. **7**(22), 1901040 (2020)
31. T. Bückmann, N. Stenger, M. Kadic, J. Kaschke, A. Frülich, T. Kennerknecht, C. Eberl, M. Thiel, M. Wegener, Tailored 3D mechanical metamaterials made by dip-in direct-laser-writing optical lithography. Adv. Mater. **24**(20), 2710–2714 (2020)
32. S. Thiele, A. Herkommer, 3D-printed microoptics by femtosecond direct laser writing, in *3D Printing of Optical Components*, ed. by A. Heinrich (Springer International Publishing, Cham, 2021), pp. 239–262
33. M. Fateri, A. Gebhardt, Introduction to additive manufacturing, in *3D Printing of Optical Components*, ed. by A. Heinrich (Springer International Publishing, Cham, 2021), pp. 1–22
34. Z. Chen, Pixelligent zirconia nano-crystals for OLED applications, in *White Paper* (2014)
35. D. Russel, A. Stabell, Scaling-up pixelligent nanocrystal dispersions, in *White Paper* (2016)
36. Z. Chen, J. Wang, Pixelligent internal light extraction layer for OLED lighting, in *White Paper* (2014)
37. Schott, Optical Glass 2020. Technical report Schott AG (2020)

Chapter 5
Achromatic Diffractive Optical Elements (DOEs) for Broadband Applications

As discussed in the introduction and Sect. 2.4, the integration of diffractive optical elements (DOEs) into a broadband optical system can often allow for increasing the system's performance, reducing its size, or its complexity. However, despite considerable efforts to develop different technologies for DOEs, they still remain highly underutilized in broadband imaging system. This is because DOEs that maintain high diffraction efficiencies across the full range of wavelengths, angles of incidence (AOIs), and grating periods required for different optical systems are currently not available (see Sect. 2.4). Since the wavelength dependence of the efficiency (see (2.16)) is fundamentally linked to the dispersion of the phase delay ($\phi(\lambda)$), this leads to the question of whether the dispersion engineering capabilities of nanocomposites could make such materials an enabling technology for finally unlocking the full potential of DOEs for optical design. In this chapter, I address this question as my first advanced application for nanocomposites. At the same time, my second goal in this chapter is to not restrict myself to one material platform and embodiment of DOEs, but also develop general concepts for how DOEs for broadband systems can be designed. The main findings presented in this chapter have already been published in different articles. The findings of Sect. 5.1 were published in Optica [1], those of Sect. 5.2 in the Journal of Optics [2], and those of Sects. 5.3 and 5.4 in Optics Express [3].

5.1 Nanocomposite-Enabled DOEs

Equation (2.16) demonstrates that for DOEs, which operate by varying the incident wave's local phase, the key to achieving high efficencies across a wide spectral range is finding a way to eliminate the wavelength dependence of their phase profiles. For

D. Werdehausen, *Nanocomposites as Next-Generation Optical Materials*, Springer Series in Materials Science 316, https://doi.org/10.1007/978-3-030-75684-0_5

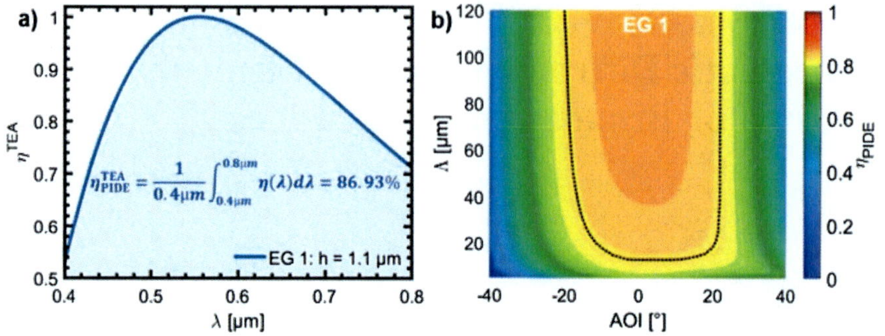

Fig. 5.1 Efficiency of a single-layer échelette-type grating made of PMMA (EG 1 in Table 5.1) from the thin element approximation (η^{TEA}) over the wavelength. The average efficiency across the shaded area corresponds to the polychromatic integral diffraction efficiency (η_{PIDE}; see (5.1)). (b) η_{PIDE} of EG 1 from finite element method (FEM) simulations (JCMsuite) over the angle of incidence (AOI) and the grating period (Λ). It's evident that η_{PIDE} decreases with decreasing grating periods (Λ) and increasing AOIs because of shadowing (see Sect. 2.4.1). The dotted black line highlights the region in which η_{PIDE} remains within 3% of its theoretical limit, which is given by η^{TEA}_{PIDE}. Reprinted with permission from [1] ©The Optical Society

échelette-type gratings (EGs), the relationship $\phi_{max}(\lambda) = \frac{2\pi h}{\lambda}\Delta n(\lambda)$ (2.17) shows that this can be achieved by tailoring the refractive index difference $\Delta n = n_1(\lambda) - n_2(\lambda)$ such that it increases linearly with the wavelength. The EG's height (h) then remains as a free parameter that can be adjusted to achieve $\phi_{max}(\lambda) = 2\pi$ and hence an efficiency of 100% for the first diffraction order ($q = 1$). As a single measure for a DOE's overall performance in the visible spectral range, I here use the polychromatic integral diffraction efficiency (η_{PIDE}), which corresponds to the average efficiency:

$$\eta_{PIDE} = \frac{1}{\lambda_2 - \lambda_1}\int_{\lambda_1}^{\lambda_2} \eta_{q=1}(\lambda)\mathrm{d}\lambda, \tag{5.1}$$

where $\eta_1(\lambda)$ is determined either from the thin element approximation, which I always indicate by the superscript TEA or numerical simulations. In the following, I assume λ_1 =400 nm and λ_2 =800 nm in (5.1), since my focus is on conventional imaging systems that operate in and around the visible spectrum. Furthermore, I focus on the first diffraction order ($q = 1$) because devices designed to operate in the first order have the minimum possible height (see (2.18)).

To first demonstrate the general problems that arise if single-layer EGs are used in broadband systems, Fig. 5.1a depicts the efficiency of a single-layer EG made from PMMA (EG 1) as a function of the wavelength. This figure illustrates that, for such a single layer structure, an efficiency of $\eta^{TEA} = 100\%$ ($\phi^{TEA}_{max}(\lambda) = 2\pi$) can only be achieved for a single wavelength (λ_0). For wavelengths other than λ_0, the efficiency decreases quickly. This is because the second layer of a single layer EG is made of air with a refractive index of $n_2(\lambda) \approx 1$ (see Fig. 2.4). It is therefore

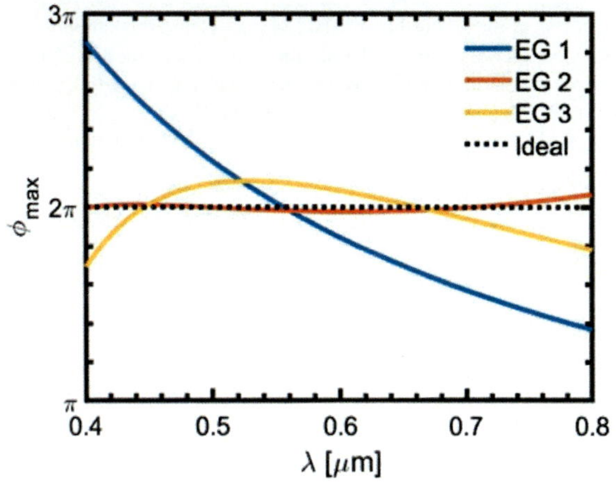

Fig. 5.2 Maximum phase delay per grating period ($\phi_{max}(\lambda)$) for EG 1 - EG 3 over the wavelength (see (2.14)). For the single layer EG 1, $\phi_{max}(\lambda)$ strongly depends on the wavelength and only takes a value of 2π for a single design wavelength (λ_0). In contrast, the phase profile of EG 2 ($\phi_{max}(\lambda)$) exhibits almost no dispersion. The phase profile of EG 3 fulfills the condition $\phi_{max}(\lambda) = 2\pi$ at two wavelengths. Reprinted with permission from [1] ©The Optical Society

impossible to find an $n_1(\lambda)$ for which $\Delta n(\lambda)$ increases close to linearly with the wavelength, since the refractive index profiles ($n(\lambda)$) of all materials with a non-unity refractive index exhibit an intrinsic curvature (see Sect. 2.1). As illustrated in Fig. 5.2, the wavelength dependence of $\Delta n(\lambda)$ hence causes $\phi_{max}^{TEA}(\lambda)$ (blue line) to quickly deviate from the optimal wavelength independent behavior (dotted black line). In fact, for the design of EG 1, the design wavelength (λ_0 in (2.17)) was optimized such that η_{PIDE}^{TEA} is maximized. This illustrates that higher average efficiencies than the one achieved for EG 1 ($\eta_{PIDE}^{TEA} = 87\%$; see Fig. 5.1) cannot be reached with a single-layer structure made from PMMA. Finally, as already discussed in Sect. 2.4.1, the efficiencies obtained from the TEA only denote the theoretical limits, which are achieved for large grating periods and small AOIs. To determine the EG's real-world performance, I therefore also performed full wave optical simulations of Maxwell's equations using the finite element method (FEM) software JCMsuite. To fully investigate the influence of the AOI and the grating period on the efficiency, I performed FEM simulations for 47 periods between 5 μm and 120 μm, as well as 41 AOIs between $-40°$ and $40°$. This range of grating periods corresponds to diffraction angles between $0.3° \leq \theta_{dif,q=1} \leq 6.7°$ for normal incidence ($\theta_{AOI} = 0$), a wavelength of $\lambda_d = 0.588$ μm, and the first diffraction order $q = 1$ (see (2.15)). For each parameter combination, I then numerically approximated the η_{PIDE} by running different simulations for 17 wavelengths between 400 nm and 800 nm. Moreover, as a first step, I assumed TM-polarization. However, in the appendix (Sect. A.2), I show that there are only negligible differences between TM- and TE-polarization. Finally, I assumed the EG's materials to extend to infinity.

Table 5.1 Specifications of the single-layer (SL) EG (EG 1) and the nanocomposite-enabled (NE) EGs (EG 2 and EG 3). Materials 1 and 2 denote the materials in the EGs' first and second layer, respectively (dia. is short for diamond; all other abbreviations are defined in Fig. 5.3). The percentages denote the respective volume fractions (f), h the EGs' heights and $\eta_{\mathrm{PIDE}}^{\mathrm{TEA}}$ the theoretical limit for the polychromatic integral diffraction efficiency that is obtained from the thin element approximation (TEA).

Name	Type	Material 1	Material 2	h (μm)	$\eta_{\mathrm{PIDE}}^{\mathrm{TEA}}$
EG 1	SL	PMMA	Air	1.1	86.93%
EG 2	NE	34.3% dia. in PMMA	35% ITO in PMMA	3.5	99.96%
EG 3	NE	21.0% ZrO_2 in PMMA	PC	18.3	98.90%

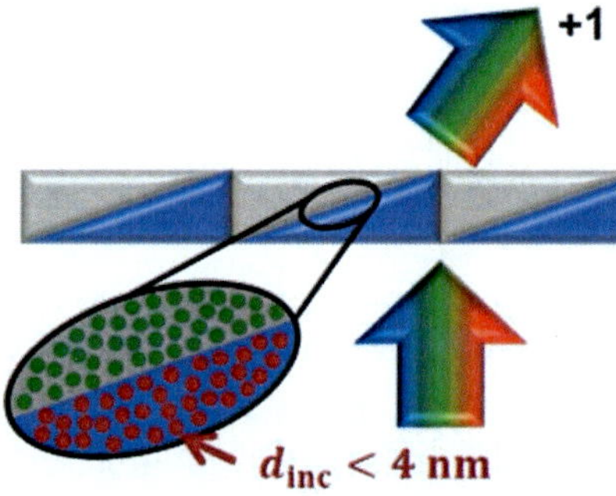

Fig. 5.3 Scheme of a nanocomposite-enabled (NE) EG that directs all incident light into the required diffraction order (here $q = 1$). Both layers are made from dispersion-engineered nanocomposites and their compositions (the refractive indices of the two layers $n_1(\lambda)$ and $n_2(\lambda)$) are optimized for a maximum diffraction efficiency (η_{PIDE}). To reach the homogeneous regime, the size of the particles (d_{inc}) must be below 4 nm to avoid incoherent scattering. Adapted with permission from [1] ©The Optical Society

Figure 5.1b depicts the η_{PIDE} of EG 1, which I obtained from the FEM simulations, as a function of the AOI and the grating period. The figure demonstrates that this EG's η_{PIDE} decreases with decreasing grating periods and increasing AOIs because of shadowing (see Sect. 2.4.1 for a detailed discussion about how I use the term shadowing). To highlight the region in which shadowing effects only play a minor role, the dotted black line visualizes the boundary of the region within which the η_{PIDE} remains within a range of 3% of the theoretical limit obtained from the TEA (see Table 5.1). The dotted line shows that this region encompasses AOIs between $-20^\circ < \mathrm{AOI} < 20^\circ$, and grating periods of $\Lambda > 15\,\mu$m. The large size of this region can be attributed to this EG's small height ($h = 1.1\,\mu$m) and the associated reduction of shadowing (see Sect. 2.4.1). However, the FEM simulations also confirm that this EG's η_{PIDE} remains below 87% even for perpendicular incidence and large grating periods.

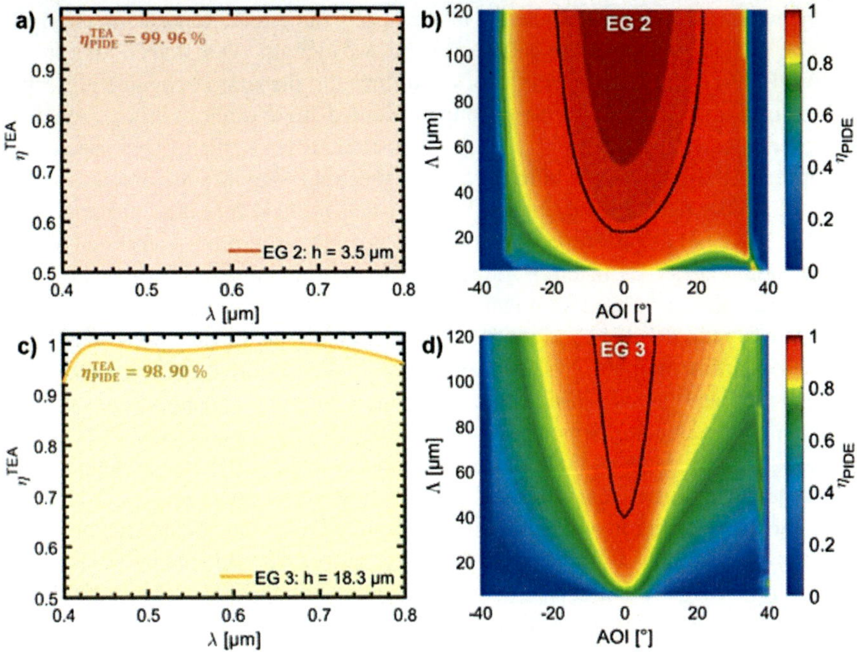

Fig. 5.4 **a** Efficiency of EG 2 (specifications in Table 5.1) from the TEA over the wavelength. The efficiency remains very close to 100% throughout the entire visible spectrum. **b** η_{PIDE} of EG 2 from finite element method (FEM) simulations (JCMsuite) as a function of the grating period (Λ) and the angle of incidence (AOI). The dotted black line indicates the region within which the η_{PIDE} stays within 3% of its theoretical limit. **c, d** Identical plots for EG 3 (specifications in Table 5.1). This EG is only made of commercially available materials. The FEM simulations for this EG in (**d**) show that both the overall magnitude of the η_{PIDE} as well as the range of AOIs and grating periods within which a high performance is maintained are reduced compared to EG 2. Reprinted with permission from [1] ©The Optical Society

5.1.1 Using the Materials as Degrees of Freedom

This finding that the η_{PIDE} of single layer EGs generally remains below 90% shows that such devices are not suited for broadband imaging applications. To investigate whether nanocomposites allow for the design of EGs with achromatic phase profiles and hence overcome the problem of low broadband efficiencies, I now assume a two-layer EG (Fig. 5.3). For the design of this device, I treated both materials as tunable nanocomposites and optimized their composition for a maximum η_{PIDE}^{TEA} and a minimum height (h). I performed this optimization using the TEA, but included the minimization of h to reduce the amount of shadowing. Moreover, I included the entire range of nanocomposites included in Fig. 4.1. In doing so, I treated the materials (host and nanoparticles) as discrete and the volume fractions as continuous parameters.

Figure 5.4a depicts the efficiency of the optimized nanocomposite-enabled EG (EG 2), as obtained from the TEA, over the wavelength. The shaded area in the figure visualizes that my approach of exploiting the dispersion-engineering capabilities of nanocomposites led to a unprecedented high value of $\eta_{\mathrm{PIDE}}^{\mathrm{TEA}} > 99.9\%$. As schematically shown in Fig. 5.3, this nanocomposite-enabled EG consequently directs essentially all light within the visible spectral range into the first diffraction order. To show that this can indeed be attributed to an achromatic phase profile, Fig. 5.2 (red line) demonstrates that this EG's $\phi_{\mathrm{max}}^{\mathrm{TEA}}(\lambda)$ is almost independent of the wavelength. Finally, as listed in Table 5.1, the optimization yielded that ITO and diamond are the best suited filler materials for the different layers. In fact, nanocomposites with both diamond and ITO as the nanoparticle materials have already been realized experimentally [4, 5]. However, future work is required towards developing transparent diamond nanoparticles, because single-digit nanodiamonds mostly have a prominent black color, which is probably caused by graphitic layers on their surfaces [6]. Finally, to show that this nanocomposite-enabled EG also maintains very high efficiencies outside the validity of the TEA, Fig. 5.4b depicts its η_{PIDE} from FEM simulations as a function of the grating period and the AOI. This figure demonstrates that this EG maintains a very high performance across a broad area in the diagram. In fact, the dotted black line again highlights that the region within which this EG's η_{PIDE} remains within 3% of its theoretical limit ($\eta_{\mathrm{PIDE}}^{\mathrm{TEA}}$) is only slightly narrower than that of the single-layer EG 1 and extends down to a grating period of $\Lambda = 22\ \mu\mathrm{m}$. In Sect. 5.2.4, I show that this region fully covers the parameter range that is required for most optical systems.

The ideal nanocomposite-enabled EG that I discussed so far (EG 2) relies on materials that have already been synthesized in research works [4, 5] but are not readily commercially available. In contrast, as discussed in the introduction, nanocomposites based on ZrO_2 have recently become commercially available [7–9]. To investigate whether high efficiencies can also be achieved using this readily available material, I therefore performed a second optimization for which I neglected all other nanoparticle materials and used a conventional polymer as the base layer. This optimization lead to the specifications of EG 3 listed in Table 5.1. Figure 5.4c shows that this EG is still characterized by a high $\eta_{\mathrm{PIDE}}^{\mathrm{TEA}}$ of 98.90%, but also an increased height of 18.3 μm. The orange line in Fig. 5.2 illustrates that this is due to a slightly more pronounced wavelength dependence of $\phi_{\mathrm{max}}^{\mathrm{TEA}}(\lambda)$. Finally, the results of the FEM simulations depicted in Fig. 5.4d show that the increased height compared to EG 2 indeed increases the amount of shadowing. This limits the range of grating periods and AOIs across which EG 3 maintains a high performance. However, in the appendix (Sect. A.3), I show that this EG still outperforms the state-of-the-art solutions for achieving efficiency achromatization. This demonstrates that nanocomposite-enabled EGs can indeed provide unprecedented high efficiencies.

5.2 Diffractive Lenses in High-Numerical-Aperture Broadband Imaging Systems

Diffractive lenses (DLs) are DOEs that implement the function of a lens. As discussed in Sect. 2.4.2, DLs are powerful tools for correcting chromatic aberrations in broadband imaging systems [10]. This is because of their highly anomalous dispersion properties, which are quantified by their effective Abbe number ($\nu_{\mathrm{d}}^{\mathrm{dif}}$) and partial dispersion ($P_{\mathrm{g,F}}^{\mathrm{dif}}$) (see Sect. 2.4.2). However, high-quality optical imaging systems commonly have contrast requirements in the order of 1/100 with regards to stray light, which implies that merely 1% of all light is allowed to propagate in spurious diffraction orders and reach the image plane. This imposes high efficiency requirements that have been almost impossible to fulfill until now. In the previous sections, I have demonstrated that nanocomposite-enabled EGs can provide unprecedented high efficiencies of essentially 100% throughout the entire visible spectrum as well as a wide range of AOIs and grating periods. But the FEM simulations for the ideal nanocomposite-enabled EG (EG 2; see Fig. 5.4b) have also shown that for grating periods around 20 µm, its average efficiency in the visible spectrum drops below $\eta_{\mathrm{PIDE}} = 97\%$ even for normal incidence. Since this grating period corresponds to a deflection angle of merely 1.69^o for an AOI of 0° and a wavelength of $\lambda = 0.588$ µm, this demonstrates that this device cannot maintain efficiencies of close to 100% for large steering angles. Therefore, a key remaining question is whether the concept of nanocomposite-enabled EGs is suitable for high-numerical-aperture (high-NA) broadband imaging systems for which much larger overall steering angles are required. To answer this question, the requirements on DLs in broadband imaging systems must be analyzed using a holistic all-system analysis that connects the system's specifications to those of the DLs. In fact, such an analysis is also required to answer the question of what technology for DLs is best suited for broadband imaging systems. This is an urgent issue, since, as discussed in the introduction, several different technologies for DLs including metalenses and multilevel DOEs are currently receiving a high amount of research interest [11–24]. In this context, the critical question is not if a DL with a high-numerical-aperture ($\mathrm{NA}^{\mathrm{dif}}$) can be realized, but rather whether the DL can be used to enhance a high-NA system without limiting its performance or imposing severe constraints on the system's design.

In this section, I provide an all-system analysis to quantify the requirements on DLs in broadband systems. To this end, I first quantify the performance of full macroscopic nanocomposite-enabled DLs. As the key performance measure for this analysis, I define a DL's focusing efficiency as the fraction of the incident power that propagates in the desired diffraction order and is hence focused into the desired focal point. Subsequently, I quantify the requirements on DLs in broadband imaging systems and investigate whether they can be fulfilled by nanocomposite-enabled DLs. This section is based on a paper that was published in the Journal of Optics [2].

5.2.1 Optical and DOE Design Perspectives

As the basis of the full-system analysis provided below, I here briefly discuss what steps are required to integrate DLs into a real-world optical systems.

First, the lens must be designed together with the overall system. In this stage, the DL is usually treated as a perfect phase plate, whose phase profile (see (2.19)) is then optimized in the design process. To this end, the dispersion properties of a DL can be described by an effective Abbe number of $\nu_d^{dif} = -3.452$ and partial dispersion of $P_{g,F}^{dif} = 0.2956$ (see Sect. 2.4.2). However, this "optical design perspective" assumes an efficiency of 100% and does not consider how this phase profile can be implemented physically. Therefore, after the optimal phase profile is determined, the key task for the "DOE-design" is to find real devices that get as close as possible to the ideal phase profiles.

While the "optical design perspective" accounts for all elements in an optical system, studies into DLs mostly neglect the fact that multiple optical elements are required to achieve high quality imaging at higher NAs and a wide field of view (FOV) [25, 26]. In fact, I emphasize that a single DL that provides aberration corrected imaging of finite-size objects, even for only a single wavelength, cannot exist [26–28]. This is because of a violation of the Abbe sine condition, which must be fulfilled to achieve well-corrected imaging close to the optical axis [26]. Furthermore, a DL is generally distinguished by fixed anomalous values of ν_d^{dif} and $P_{g,F}^{dif}$. The correction of chromatic aberrations in a hybrid system, which contains at least one DL, consequently requires both diffractive and refractive elements. In this context, I briefly note that recently multilevel DOEs and metalenses with different values of ν_d and $P_{g,F}$ have been demonstrated [21, 24, 29–33]. However, until now, the efficiencies of these DLs are way below the values that are required to address the contrast requirements of high-quality imaging systems (see [13] for an overview over what efficiencies have been achieved so far). In fact, it appears likely that efficiencies of close to 100% with these approaches will remain an insurmountable challenge. This is because both metalenses and multilevel DOEs rely on a discretization of the ideal phase profiles. It is well known that this already reduces the theoretical limit for the efficiencies below 100% [20]. Furthermore, to tailor ν_d and $P_{g,F}$, the DL's phase profile must be precisely and continuously controlled for each wavelength individually. I discuss these issues in more detail in the appendix (Sect. A.4). For broadband imaging systems, the currently most promising approach consequently is to leverage the unique values of ν_d^{dif} and $P_{g,F}^{dif}$ for the correction of chromatic aberrations in hybrid systems, that is, systems that are comprised of both refractive and diffractive elements. This is also the approach that has been shown to have a significant benefit in optical design studies [10, 34–40].

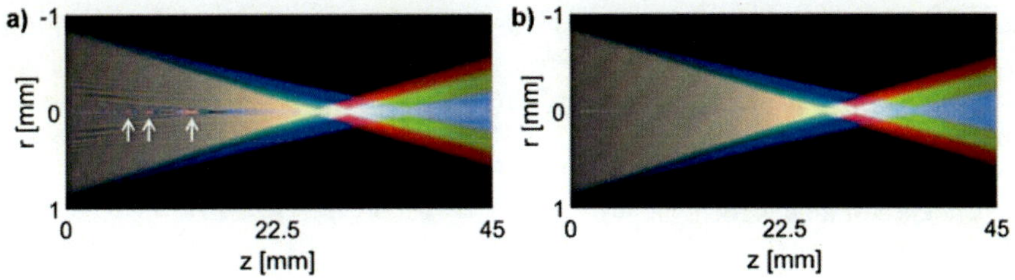

Fig. 5.5 RGB images visualizing the light propagation behind **a** a single-layer DL (based on EG 1) and **b** a nanocomposite-enabled DL (based on EG 2) at a focal length of $f = 30$ mm and an aperture of $D = 2$ mm (specifications of EGs in Table 5.1). For each of the three color channels, an independent simulation using the wave propagation method (WPM) was performed. To this end, three different wavelengths were used and the logarithms of the resulting intensity distributions were applied to the corresponding color channel (red—$\lambda_C = 656.28\ nm$, green—$\lambda_d = 587.56\ nm$, and blue—$\lambda_F = 486.13\ nm$). The arrows in (**a**) denote the positions of the 2nd, 3rd, and 4th order foci for λ_C. Reproduced from [2]

5.2.2 *Performance of Macroscopic Diffractive Lenses*

Figure 2.5 in Sect. 2.4.2 visualizes that a DL consists of segments with varying widths. Therefore, the FEM simulations for infinitely periodic structures cannot be directly applied to full DLs. In fact, because of the high requirements of full wave optical simulations on memory and computational power, the FEM is not suited for investigating the performance of full macroscopic DLs. For this purpose, I now use the recently developed rotationally symmetric formulation of the wave propagation method (WPM) [41]. This method has been shown to be a highly accurate and efficient approach for determining the focusing efficiencies of DLs based on EGs [41]. I here use the same Matlab implementation of this method that was also used in [41].

To first directly visualize the superior performance of a DL based on the ideal nanocomposite-enabled EG (EG 2; see Table 5.1) over one based on the single-layer EG (EG 1), Fig. 5.5 visualizes the propagation of polychromatic light behind DLs based on both these EGs. I obtained this image by performing an independent simulation for a wavelength in the respective spectral range for each of the three color channels in the RGB images (red ($\lambda_C = 656.28$ nm), green ($\lambda_d = 587.56$ nm), and blue ($\lambda_F = 486.13$ nm)). Moreover, I assumed a DL with a spherical phase profile, a focal length of $f(\lambda_d) = 30$ mm (see (2.19)), and an aperture of $D = 2$ mm. I then applied the logarithm of the spatial intensity distribution that I obtained from the WPM to each color channel. Furthermore, I used $E(r, z = 0) = E_0 \exp\left[-\left(\frac{2r}{\sigma_{\text{source}}}\right)^{20}\right]$, where σ_{source} corresponds to the beam's width, as the initial field. I used this function since it is almost flat across a wide range of radii around the center but quickly decays to zero around $r = \frac{\sigma_{\text{source}}}{2}$. Therefore this initial field, in good approximation, corresponds to a collimated beam of light. Finally, I used a beam width of $\sigma_{\text{source}} = 0.9D$ to make sure that the field vanishes at the edge of the lens. Comparing the images that I obtained using this procedure shows that the single-layer DL (Fig. 5.5a) exhibits prominent higher order foci (white arrows), whereas such spurious foci are not visible for the nanocomposite-enabled DL (Fig. 5.5b). In addition, the images in Fig. 5.5 confirm that within each of these foci, the DLs indeed have different focal lengths for different wavelengths. As discussed in Sect. 2.4.2, this can be attributed to the wavelength dependence of a DL's focal length.

5.2.3 Focusing Efficiency

Figure 5.5a visualizes that spurious foci are clearly visible for the single-layer DL. If light from these foci reaches the sensor, this will lead to a loss of contrast, colorful flares, and ghost images in imaging applications. Therefore, the key performance measure for DLs is the focusing efficiency (η^{foc}).

Procedure for determining the focusing efficiency
In practice, the focusing efficiency is a measure that is difficult to determine precisely. This is because the intensity patterns are always infinitely extended. Furthermore, the width and shape of the intensity pattern are functions of the wavelength and the lens' specifications (f and D). This can be attributed to aberrations and the wavelength dependence of the diffraction limit ($d_{\text{lim}} = \frac{\lambda}{2\text{NA}}$). To account for these influences, I determined the η^{foc} using the method visualized in Fig. 5.6a: (1) I fit a Gaussian to intensity pattern in the focus plane ($I(r)$) and then (2) determine the integrated intensity within the focus. To this end, I define the focus' radius in terms of the full e^{-2}-width of the Gaussian (σ_{Gauss}) according to $r^{\text{foc}} = c_{\text{foc}}\sigma_{\text{Gauss}}$, where c_{foc} is a fixed factor that I determine later. Finally, (3) I divide this result by the overall integrated intensity directly behind the lens. The inset in Fig. 5.6a illustrates that the focusing efficiency is always below 100% for this procedure. This is because, the intensity pattern is always infinitely extended even in the diffraction limit. As a benchmark, I therefore performed a simulation for a refractive lens with a focal length of $f = 30$ mm. I chose this benchmark because a refractive lens has no higher order foci. This enables me to determine a theoretical limit for the focusing efficiency, which is achieved in the absence of spurious diffraction orders. To obtain a theoretical limit that is close to 100%, I then determined the factor c_{foc} in the definition of r^{foc} from the condition that a value of $\eta^{\text{foc}} = 99\%$ is achieved for the refractive lens. This condition led to $c_{\text{foc}} = 8$. The advantage of using this fitting procedure over of a fixed value of f^{foc}, which is the approach that was used in [41], is that the fitting procedure accounts for broadening of the focal spot due to aberrations and wavelength changes. This makes the values that are determined from this procedure almost independent of the broadening of the intensity pattern caused by these influences.

Performance of macroscopic diffractive lenses
To first investigate the wavelength dependence of the DLs' focusing efficiencies, I determined both the single-layer and the nanocomposite-enabled DLs' η^{foc} at focal lengths of $f = 30$ mm and $f = 4$ mm. Accordingly, Fig. 5.6b shows that

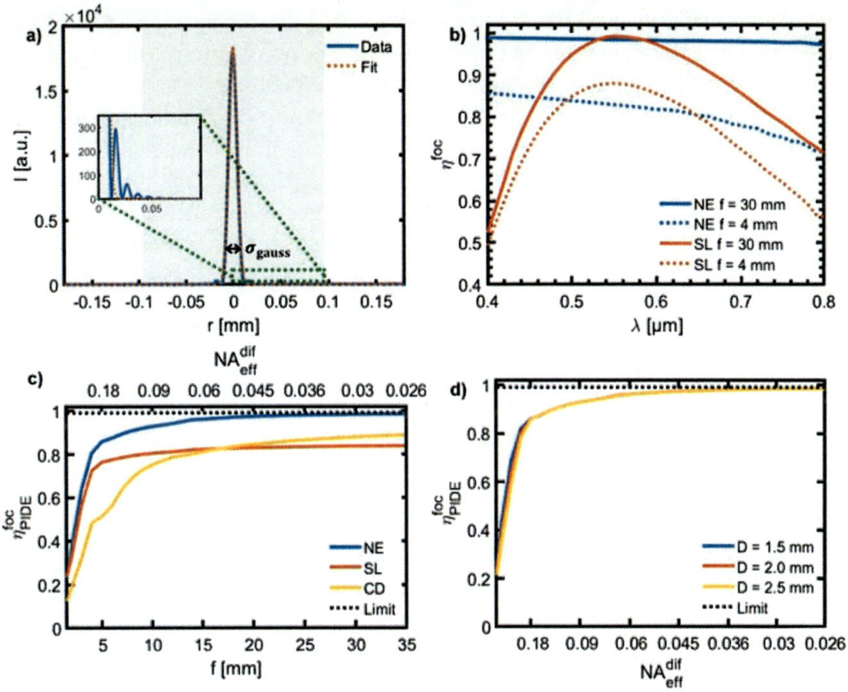

Fig. 5.6 **a** Procedure used for numerically determining the focusing efficiency (η^{foc}): A Gaussian function with a full e^{-2}-width of σ_{Gauss} was fitted to the intensity pattern in the focus plane ($I(r)$) to determine the focus spot's radius ($r^{\text{foc}} = 8\sigma_{\text{Gauss}}$). Subsequently the integrated intensity within $0 \leq r \leq r^{\text{foc}}$ (grey area) was determined and divided by the integrated intensity directly behind the lens. **b** η^{foc} over the wavelength for the single-layer (SL) DL and the nanocomposite-enabled (NE) DL at two different focal lengths and a fixed aperture of $D = 2$ mm (details of devices in Table 5.1). **c** Average focusing efficiency across the visible spectrum ($\eta^{\text{foc}}_{\text{PIDE}}$) over the focal length for both DLs. Additionally, a DL that is based on the state-of-the-art common depth (CD) EG without dispersion-engineered materials (EG S2 in Table A.1) is included (see Table A.1 in the appendix). **d** $\eta^{\text{foc}}_{\text{PIDE}}$ of the nanocomposite-enabled DL as a function of $\text{NA}^{\text{dif}} = 2\frac{f}{D}$ for different apertures. Reproduced from [2]

the nanocomposite-enabled DL indeed exhibits an almost wavelength independent focusing efficiency for the larger of the two focal lengths. In contrast, the single-layer device's efficiency drops quickly towards the edges of the visible spectral range. In fact, at this focal length, the efficiency curves are very similar to those obtained from the TEA for the corresponding infinitely periodic structures (Figs. 5.1a and 5.4a). However, the dotted lines in Fig. 5.6b show that this is no longer the case at the shorter focal length of $f = 4$ mm. At this focal length, the focusing efficiencies of both DLs drop by more than 10 percentage points because of shadowing. But the nanocomposite-enabled DL is still distinguished by a relatively flat efficiency profile and consequently still outperforms the single-layer DL. To quantify the performance of the different devices with a single measure, I now, in analogy to (5.1), define the polychromatic integral focusing efficiency ($\eta_{\text{PIDE}}^{\text{foc}}$) as the average focusing efficiency in the visible spectrum. In the following, I always determine this quantity by numerically sampling the wavelength in 50 nm increments.

To systematically quantify the focusing efficiency as a function of the focal length, Fig. 5.6c depicts the single-layer and nanocomposite-enabled DLs' $\eta_{\text{PIDE}}^{\text{foc}}$ over the focal length (f) at a fixed aperture of $D = 2$ mm. This figure confirms that the nanocomposite-enabled DL's $\eta_{\text{PIDE}}^{\text{foc}}$ converges to the theoretical limit that was obtained for the refractive lens (dotted black line) for large focal lengths. The nanocomposite-enabled DL can hence indeed provide essentially perfect broadband focusing for large focal lengths. However, it is also evident from Fig. 5.6c that, for focal lengths below $f = 15$ mm, the efficiency quickly drops below 95%. In contrast, the single-layer DL's $\eta_{\text{PIDE}}^{\text{foc}}$ converges to a much lower value ($\eta_{\text{PIDE}}^{\text{foc}} \approx 86\%$) but remains closer to this limit for short focal lengths. Finally, to demonstrate that a DL that is composed of only conventional materials cannot match the performance of the nanocomposite-enabled DL, Fig. 5.6c also presents the focusing efficiency of such a state-of-the-art DL. As explained in detail in the appendix (Sect. A.3), this DL is also composed of a two layer structure but relies on conventional materials instead of dispersion-engineered nanocomposites. The yellow line in Fig. 5.6c shows that the state-of-the-art DL's $\eta_{\text{PIDE}}^{\text{foc}}$ does not even get close to the theoretical limit and remains below 90% for all focal lengths. This confirms that dispersion-engineered materials are essential for achieving a high performance.

Role of the numerical aperture (NA)

The fact that the focusing efficiency decreases with decreasing f can be readily understood from the fundamental relationship that larger diffraction angles and consequently smaller segment widths are required to decrease the focal length. This, in turn, increases the amount of shadowing. Therefore, the minimal segment width should affect the lens' performance the most and hence be a good indicator for its overall performance. To connect the minimal segment width to the lens' specficitations (f and D), I first use (2.19) to calculate the total number of segments ($N_{\text{seg}}(r)$) within a radius of r for a DL with a spherical phase profile:

$$N_{\text{seg}}(r) = \frac{|\phi(r)|}{2\pi} = \frac{r^2}{2\lambda_0\,|f(\lambda_0)|}. \tag{5.2}$$

The local width of the segments ($\Lambda_{\text{seg}}(r)$) and their spatial frequencies $k_{\text{seg}}(r)$ can now be readily obtained from the spatial derivative of $N_{\text{seg}}(r)$:

$$k_{\text{seg}}(r) = \frac{\mathrm{d}}{\mathrm{d}r}N_{\text{seg}}(r) = \frac{r}{\lambda_0\,|f(\lambda_0)|}, \quad \text{and} \quad \Lambda_{\text{seg}}(r) = \frac{1}{k_{\text{seg}}(r)} = \frac{\lambda_0\,|f(\lambda_0)|}{r}. \tag{5.3}$$

This demonstrates that the minimal segment width (Λ_{min}) can be found at the lens' edge ($r = R$):

$$\Lambda_{\min}^{\text{dif}} = \Lambda_{\text{seg}}(R) = \frac{\lambda_0 \, |f(\lambda_0)|}{r} = \frac{\lambda}{\text{NA}^{\text{dif}}}, \tag{5.4}$$

which shows that, for a certain design wavelength (λ_0), the minimal segment width is only determined by the ratio $\frac{f}{R} = \frac{2f}{D} = 2N_f^{\text{dif}} = \frac{1}{\text{NA}^{\text{dif}}}$, where N_f^{dif} denotes the f-number of the lens and I used the approximation $\text{NA} \approx \frac{1}{2N_f}$ for the NA (see [42]). Because the f-number and the NA are quantities that are usually only used for more complicated optical systems, I use the superscript "dif" to highlight if I use these quantities for a single DL. To visualize the influence of NA^{dif} on the focusing efficiency, the upper x-axis of Fig. 5.6c directly denotes the corresponding values for each focal length. For this x-axis, and in the following, I always use the subscript "eff" to indicate that I account for the fact that the incident field only illuminates 90% of the lens' physical aperture. The data for $\text{NA}_{\text{eff}}^{\text{dif}}$ in Fig. 5.6 demonstrate that the nanocomposite-enabled DL's $\eta_{\text{PIDE}}^{\text{foc}}$ begins to deviate noticeably from the theoretical limit at a NA of $\text{NA}_{\text{eff}}^{\text{dif}} = 0.03$. From (5.4), it follows that this threshold corresponds to a minimal segment width of $\Lambda_{\min}^{\text{dif}} \approx 20\ \mu\text{m}$, which is also the value for which the simulations for infinitely periodic structures predict a significant drop of the η_{PIDE} (see Fig. 5.4b).

As the next step, to confirm that a DL's focusing efficiency is indeed only determined by its $\text{NA}_{\text{eff}}^{\text{dif}}$ and not its aperture independently, Fig. 5.6d depicts the nanocomposite-enabled DL's $\eta_{\text{PIDE}}^{\text{foc}}$ over its $\text{NA}_{\text{eff}}^{\text{dif}}$ for three different aperture diameters (D). The figure directly shows that the focusing efficiency is indeed almost independent of the lens' size. Only for large values of $\text{NA}_{\text{eff}}^{\text{dif}}$, minor differences between the different apertures are observed. These differences are caused by the fitting procedure, which I use for determining the focusing efficiency, since this procedure cannot fully compensate for the increase of spherical aberration that occurs with increasing apertures. In conclusion, Fig. 5.6d consequently demonstrates that a DL's focusing efficiency, in good approximation, only depends on its $\text{NA}_{\text{eff}}^{\text{dif}}$. The threshold of $\text{NA}_{\text{eff}}^{\text{dif}} \approx 0.03$ up to which the nanocomposite-enabled DL maintains essentially perfect broadband focusing can hence be regarded as a general limit for this technology. Spurious foci only start to appear for larger values of $\text{NA}_{\text{eff}}^{\text{dif}}$. The crucial question, which I answer in the following, hence is whether this upper limit allows for the integration of nanocomposite-enabled DLs into high-quality broadband imaging systems without introducing severe constraints on the optical design.

5.2.4 Diffractive Lenses (DLs) in Broadband Imaging Systems

For the design of high-performance broadband imaging systems, it is essential to correct for chromatic aberrations. Moreover, as discussed in Sects. 5.2 and 5.2.1, it is always necessary to combine multiple optical elements to correct for both monochromatic and chromatic aberrations. Therefore, the key task is to design DLs such that they provide exactly the amount of refractive power that is required to minimize the overall system's chromatic aberrations. At the same time, it is critical to keep the amount of light in spurious diffraction orders at a level at which the contrast requirements of the overall system can be fulfilled (see Sect. 5.2.1). To investigate if nanocomposite-enabled DOEs or, in fact, other technologies are suited for broadband imaging systems, it is therefore necessary to quantify what values of $\mathrm{NA}^{\mathrm{dif}}$ and Λ_{min} are required for such systems. In this section, I address this question by deriving analytic relationships for hybrid achromatic and apochromatic systems. To this end, I build on the relationships provided in [10] and Sect. 2.1.1, where hybrid achromats and apochromats have been investigated from the "optical design perspective". As explained in Sect. 5.2.1, this perspective neglects the DLs' microscopic structure and focusing efficiencies and hence pays no attention as to how the DL can be realized physically. In addition, I present designs for different optical systems to confirm my findings.

Diffractive lenses in hybrid achromats

As discussed in Sect. 2.1.1, an achromat is a two-element system that fulfills the condition $f(\lambda_0) = f(\lambda_1)$. It is hence the simplest system that is corrected for chromatic aberration. In a hybrid achromat, the first element is a refractive lens and the second one is a diffractive lens (DL). By introducing the ratio between the refractive lens' Abbe number ($\nu_{\mathrm{d}}^{\mathrm{ref}}$) and that of the DL ($\nu_{\mathrm{d}}^{\mathrm{dif}}$) as $q^{\mathrm{acr}} = -\frac{\nu_{\mathrm{d},1}}{\nu_{\mathrm{d},2}} = -\frac{\nu_{\mathrm{d}}^{\mathrm{ref}}}{\nu_{\mathrm{d}}^{\mathrm{dif}}}$, I can now rewrite the relationships given in (2.5) as:

$$f^{\mathrm{dif}} = f^{\mathrm{acr}}(1 + q^{\mathrm{acr}}) \quad \text{and} \quad f^{\mathrm{ref}} = f^{\mathrm{acr}} \frac{1 + q^{\mathrm{acr}}}{q^{\mathrm{acr}}}. \tag{5.5}$$

If I now again use the relationship $\mathrm{NA} \approx \frac{R}{f}$, I obtain a simple relationship for the DL's and refractive lens' individual NAs:

$$\mathrm{NA}^{\mathrm{dif}} = \frac{\mathrm{NA}^{\mathrm{acr}}}{1 + q^{\mathrm{acr}}} \quad \text{and} \quad \mathrm{NA}^{\mathrm{ref}} = \mathrm{NA}^{\mathrm{acr}} \frac{q^{\mathrm{acr}}}{1 + q^{\mathrm{acr}}}, \tag{5.6}$$

which shows that $\mathrm{NA}^{\mathrm{dif}}$ is a factor of $(1 + q^{\mathrm{acr}})^{-1}$ smaller than the overall NA of the achromat ($\mathrm{NA}^{\mathrm{acr}}$). Finally, I can use (5.4) to find an analogous relationship for the DL's minimal segment width ($\Lambda_{\mathrm{min}}^{\mathrm{acr}}$):

$$\Lambda_{\mathrm{min}}^{\mathrm{acr}} = \frac{\lambda}{\mathrm{NA}^{\mathrm{acr}}}(1 + q^{\mathrm{acr}}). \tag{5.7}$$

This relationship demonstrates that the minimal grating period ($\Lambda_{\mathrm{min}}^{\mathrm{acr}}$) increases by a factor of $1 + q^{\mathrm{acr}}$ compared to that of a single DL with the same NA as the achromat ($\mathrm{NA}^{\mathrm{dif}} = \mathrm{NA}^{\mathrm{acr}}$).

To determine typical values for q^{acr} and hence $\mathrm{NA}^{\mathrm{dif}}$ as well as $\Lambda_{\mathrm{min}}^{\mathrm{acr}}$, I now use typical material choices for positive lenses in achromats. Such materials, for example, are PMMA for systems composed of polymers or N-BK7 for systems made of glasses.

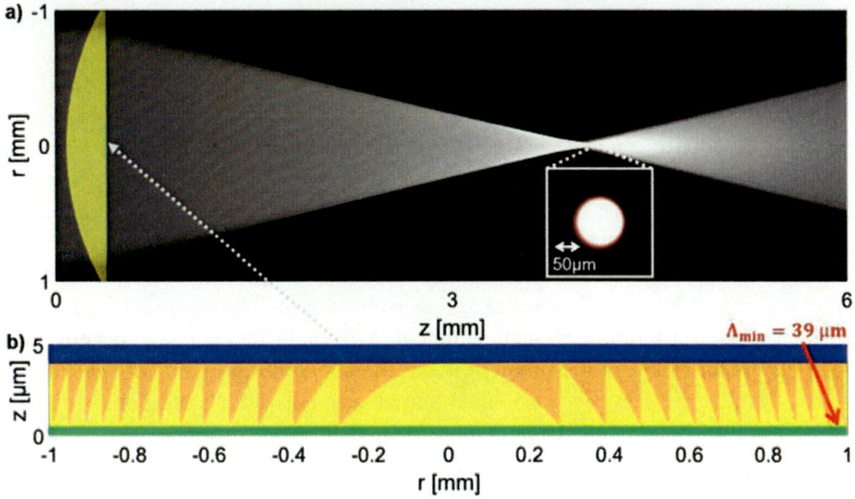

Fig. 5.7 **a** RGB image visualizing the propagation of a beam of light through a hybrid achromat with a focal length of $f^{\mathrm{acr}}(\lambda_d) = $ 4mm and a numerical aperture of $\mathrm{NA}^{\mathrm{acr}} = 0.25$. The achromat's positive lens is made of PMMA. All information on how this image was obtained can be found in the description of Fig. 5.5. The figure's inset visualizes the intensity profile in the focus plane for a heavily saturated sensor (backlighting). **b** Scheme of the nanocomposite-enabled DL (EG 2; see Table 5.1) that is located on the refractive lens' second surface. The four colors correspond to the different materials (green: PMMA; yellow: diamond in PMMA nanocomposite; orange: ITO in PMMA nanocomposite; blue: air). Reproduced from [2]

The Abbe numbers of these prototype materials are $\nu_{\mathrm{d}}^{\mathrm{PMMA}} = 58.00$ [43] and $\nu_{\mathrm{d}}^{\mathrm{BK7}} = 64.17$ [44], respectively. For these materials, the factor q^{acr} hence takes values of $q_{\mathrm{PMMA}}^{\mathrm{acr}} = 16.80$ and $q_{\mathrm{BK7}}^{\mathrm{acr}} = 18.59$. In a hybrid achromat, the DL's individual NA ($\mathrm{NA}^{\mathrm{dif}}$) is consequently more than an order of magnitude lower than the NA of the overall system ($\mathrm{NA}^{\mathrm{acr}}$). Using the limit of $\mathrm{NA}^{\mathrm{dif}} = 0.03$ up to which nanocomposite-enabled DLs can achieve almost perfect broadband focusing, now shows that this technology allows for the realization of hybrid achromats with $\mathrm{NA}_{\mathrm{PMMA}}^{\mathrm{acr}} \leq 0.53$ or even $\mathrm{NA}_{\text{N-BK7}}^{\mathrm{acr}} \leq 0.59$ for PMMA and N-BK7, respectively. This upper limit, for example, already covers the range of high-end camera lenses. This can be readily seen from the fact that these NAs correspond to f-numbers of $N_f \geq 0.94$ for PMMA and $N_f \geq 0.85$ for N-BK7. In conclusion, this analysis demonstrates that high $\mathrm{NA}^{\mathrm{dif}}$-DLs are not required for hybrid achromatic systems. This also shows that nanocomposite-enabled DLs are highly suitable for such systems. In fact, below I show that the limits on the NAs are even further relaxed in systems that are corrected for chromatic aberrations to a higher degree than achromats (see Sect. 2.1.1).

To confirm the validity of the equations derived above and investigate the performance of a prototype hybrid achromat that includes a nanocomposite-enabled DL, I designed a hybrid achromat with a focal length of $f^{\mathrm{acr}}(\lambda_{\mathrm{d}}) = 4$ mm and an aperture of $D = 2$ mm. I chose these specifications, which correspond to a NA of $NA^{acr} = 0.25$, because this NA is the upper limit for which the achromat can still achieve a diffraction-limited performance. For higher NAs, spherical aberration drastically impedes the achromat's performance. As the first step, I optimized the achromat within "OpticStudio" using ray-tracing, and then implemented the system into Matlab, where I again used the WPM to perform a wave-optical simulation. The result of this simulation is visualized in Fig. 5.7, which presents an RGB image that visualizes the propagation of light through the hybrid achromat. I again used the method described in the caption of Fig. 5.5 to obtain this image. First, the scheme of the DL's structure in Fig. 5.7b shows that it is characterized by a minimal segment width of $\Lambda_{min}^{acr} = 39\ \mu m$. This value agrees well with the one obtained from (5.4), which predicts a minimal segment width of $\Lambda_{min}^{acr} = 42\ \mu m$ for this system. The small difference between the prediction and the actual design is mostly caused by the lens' finite thickness. Because this minimal segment width corresponds to a NA of $NA^{dif} = 0.015$ (5.4), this result demonstrates that the DL is indeed operated way below the limit for which spurious diffraction orders start to appear. This can also be directly seen from the lack of spurious foci in Fig. 5.7a. Finally, to directly visualize that no spurious orders appear, even for a heavily saturated sensor, the inset in Fig. 5.7a depicts the intensity profile in the focus plane under such conditions. This figure shows that, even in this worst-case scenario of strong backlighting, no stray light is visible. There is only a slight red hue around the spot, which can be attributed to the increase of the diffraction limit that occurs with increasing wavelengths ($d_{lim} = \frac{\lambda}{2NA}$).

Efficiency requirements
The finding that DLs in hybrid systems must only provide little refractive power implies that only small values of NA^{dif} are required. However, it also entails profound consequences for the required efficiency. This is because the less refractive power a DL in a hybrid system provides, the smaller the separation between the different orders becomes. This can be readily seen from Fig. 5.8a which visualizes the propagation of light for the same achromat as Fig. 5.7 but for the single-layer DL instead of the nanocomposite-enabled one. In Fig. 5.8a, the presence of spurious foci is clearly visible as additional rays around the primary focus. In addition, the figure's inset shows that these spurious foci manifest themselves as colorful fringes in the focus plane. Note that I again assumed a heavily saturated sensor for this visualization. This highlights that the different orders propagate very close to each other. Therefore, most of the light in the spurious diffraction orders reaches the sensor. In real imaging applications, this can lead to colorful ghost images. With high accuracy, it can hence be assumed that almost all the light propagating in spurious diffraction orders reaches the sensor. In the appendix (Sect. A.5), I investigate this relationship in detail. Therefore, the contrast requirement of high-quality imaging systems ($\frac{1}{100}$) can only be fulfilled if the focusing efficiency remains within around 1% of the theoretical limit. As shown in Sect. 5.2.3, this can be achieved for nanocomposite-enabled DLs for $NA^{dif} \leq 0.03$ but not for common depth EGs with conventional materials, single-layer DLs, or current metalens designs (see [21] for an overview over what efficiencies have been achieved so far).

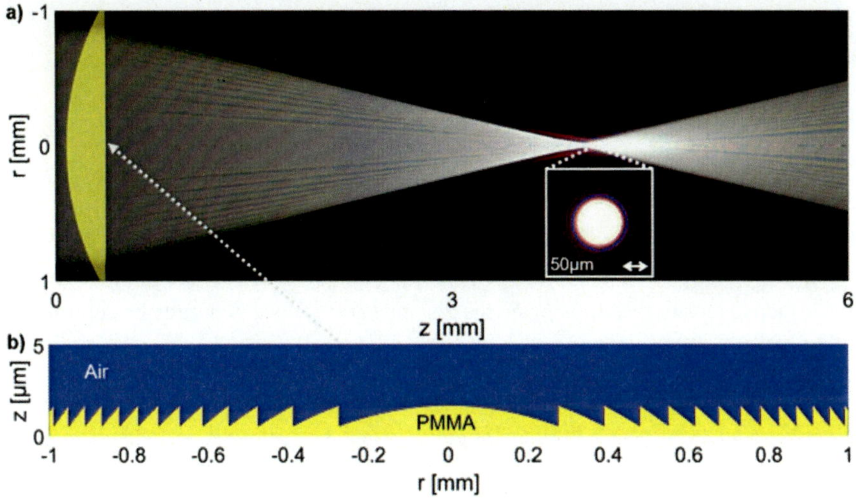

Fig. 5.8 **a** RGB image visualizing the propagation of a beam of light through a hybrid achromat with a focal length of $f^{acr}(\lambda_d) = 4$ mm and a numerical aperture of $NA^{acr} = 0.25$. The achromat's positive lens is made of PMMA. All information on how this image was obtained can be found in the description of Fig. 5.5. The figure's inset visualizes the intensity profile in the focus plane for a heavily saturated sensor (backlighting). **b** Scheme of the single-layer DL (EG 2; see Table 5.1) that is located on the refractive lens' second surface. Reproduced from [2]

Hybrid apochromats

For wavelengths other than the two specific design wavelengths, the focal lengths of achromats still deviate from the specified focal length. This remaining color error is commonly called the "secondary spectrum" and its correction is critical for high-end imaging systems. In optical design, correcting for the secondary spectrum, is usually called "apochromatization" [25]. An apochromat, in this context, is a system that fulfills the condition $f(\lambda_1) = f(\lambda_2) = f(\lambda_3)$ for three different wavelengths (λ_1, λ_2, and λ_3). In the following, I choose $\lambda_1 = \lambda_g$, $\lambda_2 = \lambda_F$, and $\lambda_3 = \lambda_C$. This choice implies that the Abbe number ν_d and partial dispersion $P_{g,F}$ as defined in (2.2) appear in all expressions that I derive below. However, by using other spectral lines for their definitions, all results can be readily generalized to other wavelengths choices.

To match the focal lengths for three different wavelengths, hybrid apochromats are comprised of one DL and two refractive lenses that are made of two different materials [10]. The elements' individual focal lengths can then be readily calculated using the relationships provided in (2.6). I now use these equations as a starting point for investigating the properties of DLs (Λ_{min} and NA^{dif}) in hybrid apochromats. To this end, I first write the linear function that connects the two refractive lenses' materials in the $P_{g,F}$-diagram (Fig. 2.6b) as $P^{ref}_{g,F}(\nu^{ref}_d) = \alpha + \beta\nu^{ref}_d$. In addition, I introduce $\Delta P^{dif}_{g,F}$ as the distance of a DL's partial dispersion ($P^{dif}_{g,F}$) from this linear function (see Fig. 2.6 for a visualization). For the DL, the corresponding relationship hence reads $P_{g,F}^{dif}(\nu_d^{dif}) = \alpha + \beta\nu_d^{dif} + \Delta P_{g,F}^{dif}$. Substituting these expressions into (2.6) and (2.7) now leads me to simple equations for f^{dif} and consequently also NA^{dif}:

$$f^{dif} = -f^{apo}\frac{\Delta P_{g,F}^{dif}}{\beta\nu_d^{dif}} = f^{apo}q^{apo} \quad \text{and} \quad NA^{dif} = \frac{NA^{apo}}{q^{apo}}, \tag{5.8}$$

where I introduced the coefficient $q^{apo} = -\frac{\Delta P_{g,F}^{dif}}{\beta\nu_d^{dif}}$. This factor is positive since the coefficients $\Delta P_{g,F}^{dif}$, β, and ν_d^{dif} are all negative. For each choice of the two materials in the refractive lenses, the slope β and the distance $\Delta P_{g,F}^{dif}$ must generally be determined individually from:

$$\beta = \frac{P_{g,F}^{ref,2} - P_{g,F}^{ref,1}}{\nu_d^{ref,2} - \nu_d^{ref,1}} \quad \text{and} \quad \Delta P_{g,F}^{dif} = P_{g,F}^{dif} - P_{g,F}^{ref,i} + \beta(\nu_d^{ref,i} - \nu_d^{dif}), \tag{5.9}$$

with $i \in \{1, 2\}$. But, for most optical glasses or polymers, the respective normal lines (see visualization in Fig. 2.6) are a good approximation. For glasses, this normal line is commonly defined as the linear function that connects the glasses K7 and F2 [45]. Finally, by substituting the expression for NA^{dif} from (5.8) into (5.4), I find that the DL's minimum segment width also increases by the factor q^{apo} compared to a diffractive singlet with the same NA as the apochromat:

$$\Lambda_{min}^{apo} = \frac{\lambda}{NA^{apo}}q^{apo}. \tag{5.10}$$

I can now directly obtain typical values for q^{apo} by using the glasses that define the normal line (K7 and F2) as a first example and, in addition, defining an analogous line for polymers as the line that connects PMMA and PC. Figure 2.6 demonstrates that this line is indeed a good approximation for the properties of a wide range of optical polymers. For the glasses, this leads to $\beta^{glass} = -0.00168$ and $\Delta P_{g,F}^{dif,glass} = -0.354$, which corresponds to $q^{apo,glass} = 59.0$. Furthermore, for the polymers, I even obtain $\beta^{poly} = -0.000507$ and $\Delta P_{g,F}^{dif,poly} = -0.293$, which results in $q^{apo,poly} = 164$. This demonstrates that the DL's NA (NA^{dif}) is around two orders of magnitude lower than the overall apochromat's NA (NA^{apo}). At the same time, the minimal segment width (Λ_{min}) increases by the same factor compared to a single DL with the same NA. The threshold of $NA^{dif} = 0.03$, up to which nanocomposite-enabled DLs can maintain essentially perfect broadband focusing, hence covers the requirements of apochromats with NAs of up to $NA^{apo} \leq 1.77$. This range encompasses essentially all existing imaging systems, even high-NA immersion objectives. Conversely, to realize a system with $NA^{apo} = 1$, which is the theoretical limit for a system operating in air, a DL with only $NA^{dif} = 0.017$ ($\Lambda_{min} = 34.7\ \mu m$) or $NA^{dif} = 0.0061$ ($\Lambda_{min} = 96.4\ \mu m$) is required for the glasses or polymers, respectively. In fact, below I demonstrate that the relationships derived above also remain a good approximation for more complex apochromatic optical systems. This finding shows that nanocomposite-enabled

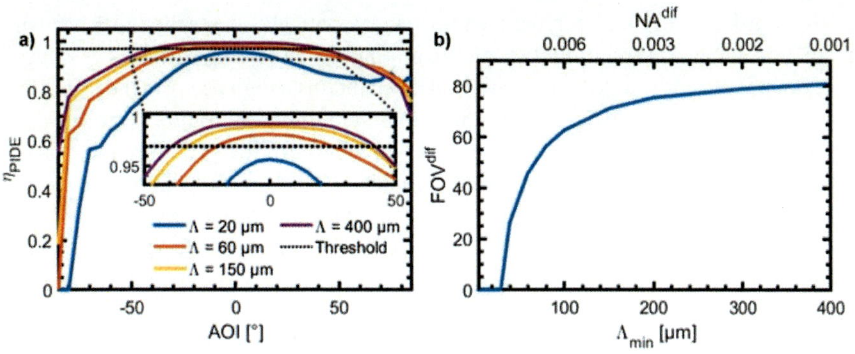

Fig. 5.9 **a** Polychromatic integral diffraction efficiency (η_{PIDE}) over the AOI for the nanocomposite-enabled EG (EG 2 in Table 5.1) assuming an infinitely periodic structure. Air ($n(\lambda) \approx 1$) was used as the surrounding medium and the data was obtained from the FEM simulations (JCMsuite). The different colors depict η_{PIDE} for four different grating periods ($\Lambda = \Lambda_{\text{min}}$) and the dotted line highlights the efficiency threshold of 97%. **b** Field of view (FOV^{dif}) for the same EG as a function of Λ_{min}. An EG's field of view (FOV^{dif}) is defined as the full range of AOIs for which the EG's η_{PIDE} remains above the efficiency threshold of 97%. The figure's upper x-axis denotes the value of NA^{dif} that corresponds to each minimal grating period (Λ_{min}) (from (5.4)). Because most segments of a DL are much larger than the DL's minimum grating period (Λ_{min}), these values estimate a lower boundary for a DL's FOV^{dif}. Reproduced from [2]

DLs in high-end broadband systems must be operated well below the threshold at which their performance starts to deteriorate. High-NA diffractive singlets are hence not required for broadband imaging systems. In fact, if a DL with a high value of NA^{dif} were to be integrated into a broadband imaging system, this would generally even decrease the system's performance. This is because such a DL would lead to highly overcorrected chromatic aberrations. Note that this situation only changes for monochromatic or very narrowband applications, for which the chromatic aberration induced by the DL does not play a role.

Field of view

In Sect. 5.2.3, I have shown that nanocomposite-enabled DLs can provide practically perfect broadband focusing up to a NA of NA$^{\text{dif}} = 0.03$. Furthermore, in the previous sections, I have analytically demonstrated that this threshold fully meets the requirements of high-end broadband imaging systems. The remaining essential requirement for the integration of DLs into broadband systems now is that the DLs must also maintain a high performance across the entire range of AOIs that is incident upon the DL. This is because a certain field of view (FOV), that is, a certain range of AOIs is always required to achieve imaging of finite sized objects. While, in general, it would also be possible to introduce constraints on the permitted range of AOIs at the DL's position, such constraints would at least partly reduce the benefits of adding the DL in the first place. To access the full potential of DL for broadband optical systems, the DLs must therefore generally also maintain high focusing efficiencies across the systems' entire FOV or an even larger range of AOIs.

Since a DL's minimal segment width ($\Lambda_{\min}$) critically determines its focusing efficiency, I now use an infinitely periodic grating with a grating period of $\Lambda = \Lambda_{\min}$ to estimate the range of AOIs for which a DL maintains a high performance. Because most of a DL's segments are much larger than its minimum grating period ($\Lambda_{\min}$), this approach allows me to estimate a lower boundary for the overall range of AOIs for which a DL maintains a high focusing efficiency. I use this procedure, because the WPM becomes inefficient for large AOIs because of the broken cylindrical symmetry. Accordingly, Fig. 5.9a depicts the polychromatic integral diffraction efficiency (η_{PIDE}) for the nanocomposite-enabled EG (EG 2 in Table 5.1) over the AOI. I again obtained these data using the FEM software JCMsuite. Moreover, for these simulations, I used air with $n = 1$ as the surrounding medium, because the layers' individual optical elements are generally placed in air or vacuum in most optical systems. Note that in Sect. 5.1 (Fig. 5.4), I performed the FEM simulations under the assumption that the EG's materials are extended to infinity. This leads to lower efficiencies for large AOIs because the incident light is refracted towards the normal line at the additional interfaces if the EGs are placed in air. Figure 5.9a demonstrates that the nanocomposite-enabled EG's η_{PIDE} remains almost constant across a broad range of AOIs for large grating periods (see also Fig. 5.4). In contrast, for shorter grating periods, the efficiencies decrease more rapidly with increasing AOIs. To quantify this effect systematically, I define the EG's field of view (FOV^{dif}) as the entire range of AOIs across which the η_{PIDE} remains above 97%. This threshold is visualized by the dotted black line in Fig. 5.9a. Using this definition, Fig. 5.9b demonstrates that the EG's FOV^{dif} indeed decreases quickly with decreasing grating periods. However, most importantly, for $\Lambda_{\min} \geq 200$ μm ($\text{NA}^{\text{dif}} \leq 0.003$), a FOV of $\text{FOV}^{\text{dif}} > 75°$ is achieved and, for $\Lambda_{\min} \geq 100$ μm ($\text{NA}^{\text{dif}} \leq 0.006$), still a FOV of $\text{FOV}^{\text{dif}} > 62°$. These values exceed the FOVs of a wide range of different imaging systems by far. Specifically, microscopes mostly have FOVs below 8°. In contrast, photographic camera lenses cover a wider range that reaches from 2° − 4° for extreme telephoto and 43° − 56° for standard lenses [25]. In conclusion, this shows that nanocomposite-enabled DLs indeed fulfill the requirements of most broadband imaging systems.

Prototype telephoto lens

Telephoto lenses are ideal systems for investigating the potential of hybrid systems that are composed of both refractive and diffractive lenses (DLs). This is because they suffer heavily from longitudinal chromatic aberration because of their long focal lengths. To demonstrate that nanocomposite-enabled DLs allow for unlocking the full potential of DLs for telephoto lenses with a negligible negative impact on the systems' performance or additional constraints on their design, I now investigate a prototypical apochromatic telephoto system. This, at the same time, allows me to investigate if the results from the previous sections also hold true for more complex optical systems and investigate the potential of DLs to enhance optical systems on one specific example. For these purposes, I designed a telephoto system with a focal length of $f = 100$ mm, a field of view of FOV = 10°, a f-number of $N_f = 5.6$, and a total system length of TL = 85 mm. I selected these specifications such that the optimized system, as discussed below, achieves almost diffraction-limited performance. Furthermore, I

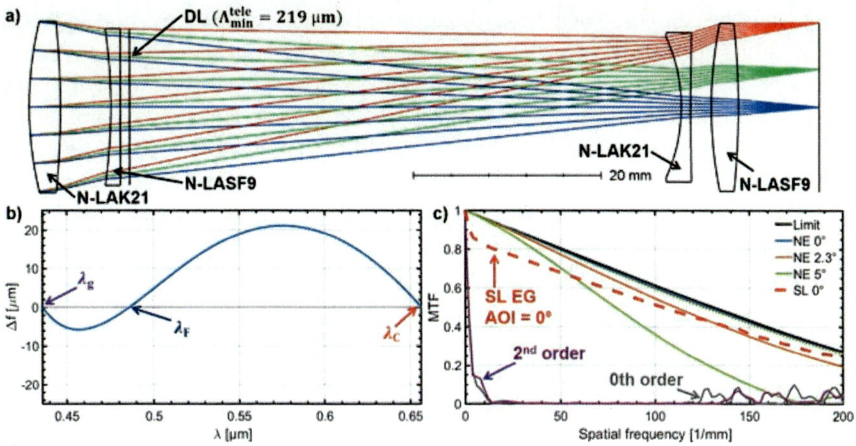

Fig. 5.10 **a** Layout of a prototypical telephoto lens. The system has a focal length of $f = 100$ mm, a field of view of FOV $= 10°$, a f-number of $N_f = 5.6$, and a total system length of TL $= 85$ mm. The three colors of the different ray bundles correspond to different field angles, that is, locations in the object plane from which the rays originate. For this design, the object was assumed to be at an infinite distance from the first lens. As highlighted in the figure, the unconstrained optimization yielded a DL with a minimal segment width of $\Lambda_{\text{min}}^{\text{tele}} = 219$ μm ($\text{NA}^{\text{dif}} = 0.0027$). This demonstrates that high-NA DLs are not required for broadband optical systems. **b** Focal shift Δf over the wavelength (λ). Its evident that the system fulfills the condition $f(\lambda_g) = f(\lambda_F) = f(\lambda_C)$. **c** Modulation transfer function (MTF) as a function of the spatial frequency for different field angles in both the sagittal (dashed lines) and tangential planes (solid lines). The grey and purple curves visualize the MTF for the dominant spurious diffraction orders (the 0th and 2nd orders). In addition, the dashed red curve demonstrates, for a field angle of 0°, that these higher order foci drastically impair the system's performance if a single layer (SL) DL is integrated into the system instead of the nanocomposite-enabled DL. Reproduced from [2]

again used "OpticStudio", which relies on ray-tracing to optimize the system. The optimized system's layout is presented in Fig. 5.10a. The differently colored rays in this figure correspond to ray bundles that originate from different points of the object, that is, different points in the object plane. Note that, in ray-tracing, the trajectory of each of these rays through the system is determined using Snell's law, which is used to calculate the angles of refraction at all interfaces. This allows for optimizing the system such that all rays originating from one point in the object plane are ideally focused onto one point in the image plane, that is, one point of the sensor. Note that, for simplicity, I assumed that the object is located at an infinite distance from the system for the optimization.

The layout of the optimized telephoto system in Fig. 5.10a illustrates that I chose a design with four refractive elements for the prototype system. I selected four lenses because a telephoto lens, that is, a lens with a focal length that is larger than its total length ($f >$ TL) must be comprised of at least a negative and a positive group [25]. To be able to correct the system for chromatic aberration, both these groups must then be realized as achromats individually [25]. Therefore, an overall achromatic telephoto system must be composed of at least four different optical elements. Second and most importantly, I added a DL as a flat additional element and introduced no constraints on this element at all. This also implies that it was allowed to freely move within the entire system. Finally, I also used the refractive lens' materials as free parameters, but constrained the second group to be made of the same materials as the first one. I introduced this constraint because it corresponds to the "worst case scenario" in which the DL must compensate for the chromatic aberration induced by both groups to achieve apochromatization. In the appendix (Sect. A.6), I additionally present a design without the DL to show that apochromatization is essential for achieving a high performance.

For the hybrid telephoto system, Fig. 5.10b depicts the optimized system's focal shift ($\Delta f(\lambda) = f(\lambda_F) - f(\lambda)$) over the wavelength to show that the system is indeed apochromatic ($f(\lambda_g) = f(\lambda_F) = f(\lambda_C)$). This wouldn't be possible without the DL because of the constraint on the system's materials (see Sect. A.6). Second, the system's modulation transfer function (MTF) in Fig. 5.10c demonstrates that the system indeed achieves almost diffraction limited performance across the full FOV. The system's performance is only slightly impaired in the tangential plane at the field angle of 5°. Note that the system's MTF is a direct measure for its performance because the MTF quantifies the system's contrast as a function of the spatial frequency in the image. Furthermore, as noted in Fig. 5.10a, the fully unconstrained optimization of the DL yielded a minimal segment width of $\Lambda_{\mathrm{min}}^{\mathrm{tele}} = 219\ \mu\mathrm{m}$, which corresponds to a NA of $\mathrm{NA}^{\mathrm{dif}} = 0.0027$. This confirms that the DL is indeed operated well below the limit of $\mathrm{NA}^{\mathrm{dif}} = 0.03$ at which the performance of the nanocomposite-enabled DL starts to be slightly impaired. In addition, Fig. 5.9 shows that for these values of $\Lambda_{\mathrm{min}}^{\mathrm{tele}}$ and $\mathrm{NA}^{\mathrm{dif}}$, the DL's field of view ($\mathrm{FOV}^{\mathrm{dif}}$) is much larger than the system's FOV of 10°. This demonstrates that spurious diffraction orders remain suppressed if the nanocomposite-enabled DL is used in the system. In fact, this result also shows that nanocomposite-enabled DLs are perfectly suited for high-end systems with higher FOVs and f-numbers. This is because the DL is operated an order of magnitude below the $\mathrm{NA}^{\mathrm{dif}}$ at which its performance starts to deteriorate.

To demonstrate that the prototype system's performance would be severely impaired if a significant amount of power propagated in the DL's spurious orders, the purple and grey curves in Fig. 5.10c visualize the MTF for the dominant spurious diffraction orders (for a field angle of 0°). This figure demonstrates that the corresponding MTF values are essentially zero for most spatial frequencies. If a significant amount of light reaches these orders, their influence consequently severely affects the system's performance. This can be directly seen from the dashed red curve, which denotes the system's MTF at a field angle of 0°, if a single-layer DL is used instead of the nanocomposite-enabled one. I obtained this curve using the data depicted in Fig. 5.6c, which show that the $\eta_{\mathrm{PIDE}}^{\mathrm{foc}}$ of the single-layer DL is approximately 15 percentage points (p.p.) below the theoretical limit for small values of $\mathrm{NA}^{\mathrm{dif}}$. The dashed red line in Fig. 5.10c demonstrates that this causes the MTF to drop by more than 10 p.p. at all spatial frequencies. This shows that the system's performance is severely impaired. Note that current metalens designs are characterized by even lower efficiencies than the single-layer DL (see summary in [21]). This shows that such devices are still far from being suitable for high-quality imaging systems. Finally, I show in the appendix (Sect. A.6) that the performance of a purely refractive system with the same specifications and the same number of elements is significantly reduced compared to that of the hybrid system presented here. This illustrates that DLs are powerful tools for optical design. I investigate this aspect in more detail in Sect. 6.3.

Finally, to investigate whether the analytic expressions for apochromats that I derived in Sect. 5.2.4 also provide a good estimate for more complex optical systems, I now assume an apochromat that is made of the the same materials (N-LAK21 and N-LASF9) and has the same NA as the prototype telephoto system (Fig. 5.10). For these materials, I obtain a factor of $q_{\mathrm{apo}} = 54.9$ and hence a minimal segment width of $\Lambda_{\mathrm{min}}^{\mathrm{apo}} = 362\ \mu\mathrm{m}$ ((5.8) and (5.10)). The finding that this value deviates by around 40% from the minimal segement width of $\Lambda_{\mathrm{min}}^{\mathrm{tele}} = 219\ \mu\mathrm{m}$ that I obtained for the telephoto lens is a consequence of the fact that the DL must work against the chromatic aberration induced by both refractive groups to achieve apochromatization. This shows that in more complex apochromatic optical systems, which are composed of more than three lenses, the optimal values of Λ_{min} might be somewhat smaller ($\mathrm{NA}^{\mathrm{dif}}$ somewhat larger) than the values obtained from (5.10). However, because much smaller values of Λ_{min}, or equivalently much larger values of $\mathrm{NA}^{\mathrm{dif}}$, would lead to highly overcorrected chromatic aberration, these quantities are generally in the same order of magnitude as the values predicted by (5.8) and (5.10). The analytic relationships from Sect. 5.2.4 can therefore provide a valuable first estimate even for more complicated apochromatic systems.

5.3 A General Design Formalism for Highly Efficient Broadband DOEs

In the previous sections, I have demonstrated that nanocomposite-enabled diffractive lenses (DLs) allow for unlocking the potential of DLs for broadband imaging systems without any negative impact on their performance from stray light or constraints on their design. In doing so, I have also shown that high-NA DLs are not required for broadband optical systems. They are rather even detrimental to their performance. Therefore, research into DLs for broadband optical systems should focus on achieving focusing efficiencies of close to 100% rather than high NAs. However, until now, I have only presented high-performance EGs that are made of specific combinations of dispersion-engineered materials. A general formalism for the design of highly efficient broadband DOEs and rules on how the dispersion must be tailored for this purpose are still missing. Such a general framework is essential to enable the large-scale integration of DLs into optical systems. This is because such a framework is required to guide the search for materials that fulfill the optical requirements and, at the same time, enable a low-cost and high-volume fabrication. Furthermore, a key question that arises from the previous sections is whether my approach of using dispersion-engineering to achieve efficiencies of close to 100% across a broad wavelength range, can also be generalized to other technologies for DOEs, for example, metalenses. In this section, I address these open challenges. This section is based on a paper that was published in Optics Express [3].

5.3.1 Design Framework

In Sect. 5.1, I have already discussed that the key to designing EGs with achromatic phase profiles and hence efficiency curves is to engineer the materials' refractive index profiles such that $\Delta n(\lambda) = n_1(\lambda) - n_2(\lambda)$ scales almost linearly with the wavelength (see (2.16) and (2.17)). This is not a straightforward task since the refractive index profiles ($n(\lambda)$) of all optical materials exhibit an intrinsic curvature (see Sect. 2.1). To generalize my approach of nanocomposite-enabled EGs, I now first develop a general framework that allows for identifying all combinations of $n_1(\lambda)$ and $n_2(\lambda)$ for which $\Delta n(\lambda)$ scales approximately linearly with λ. I then use this framework to develop rules for the design of broadband EGs and a phenomenological relationship that directly allows determining the required $n_2(\lambda)$ if $n_1(\lambda)$ is fixed. Lastly, I include the requirement that a high performance must also be maintained across a broad range of grating periods (Λ) and AOIs.

As the general basis of my framework, I describe the refractive index profiles $n_1(\lambda)$ and $n_2(\lambda)$ by their refractive indices at the d-line (n_d), Abbe numbers (ν_d), as well as partial dispersions ($P_{\mathrm{g,F}}$). Subsequently, I use Cauchy's equation ($n(\lambda) = B + \frac{C}{\lambda^2} + \frac{D}{\lambda^4}$; see Sect. 2.1) to obtain the full wavelength dependence of $n_i(\lambda)$ ($i \in \{1, 2\}$) throughout the visible spectrum for each set of n_d, ν_d, and $P_{\mathrm{g,F}}$. Since, as discussed in Sect. 2.1, Cauchy's equation is a valid approximation for transparent optical materials within the visible spectral range, this approach allows me to determine what properties of $n_1(\lambda)$ and $n_2(\lambda)$ are required independent from the materials' actual chemical composition. As a single measure for the EGs' overall performance, I then again use the polychromatic integral diffraction efficiency (η_{PIDE}), which corresponds to the average efficiency between $\lambda = 0.4$ μm and $\lambda = 0.8$ μm (see (5.1)). Finally, I use the expressions obtained from the thin element approximation (TEA; see (2.16) and (2.17)) to determine the efficiency and height for each combination of $n_1(\lambda)$ and $n_2(\lambda)$. In doing so, I again optimize the EG's height, that is, its design wavelength (λ_0) (see (2.18)) for a maximum $\eta_{\mathrm{PIDE}}^{\mathrm{TEA}}$.

Achieving achromatic phase profiles

Within my framework, finding optimal combinations of $n_1(\lambda)$ and $n_2(\lambda)$ is a six-dimensional problem ($n_{\mathrm{d},i}$, $\nu_{\mathrm{d},i}$, and $P_{\mathrm{g,F},i}$ for $i \in \{1, 2\}$). To be able to systematically investigate the influence of the individual parameters, I consequently need to reduce the dimensionality. As the first step, I therefore fix the refractive index of the first material ($n_1(\lambda)$) at $n_{\mathrm{d},1} = 1.8$ $\nu_{\mathrm{d},1} = 60$, and $P_{\mathrm{g,F},1} = 0.55$. I selected these parameters because they correspond to the upper left corner of the region that is covered by nanocomposites in the Abbe diagram (Fig. 4.1a). In addition, I fix both partial

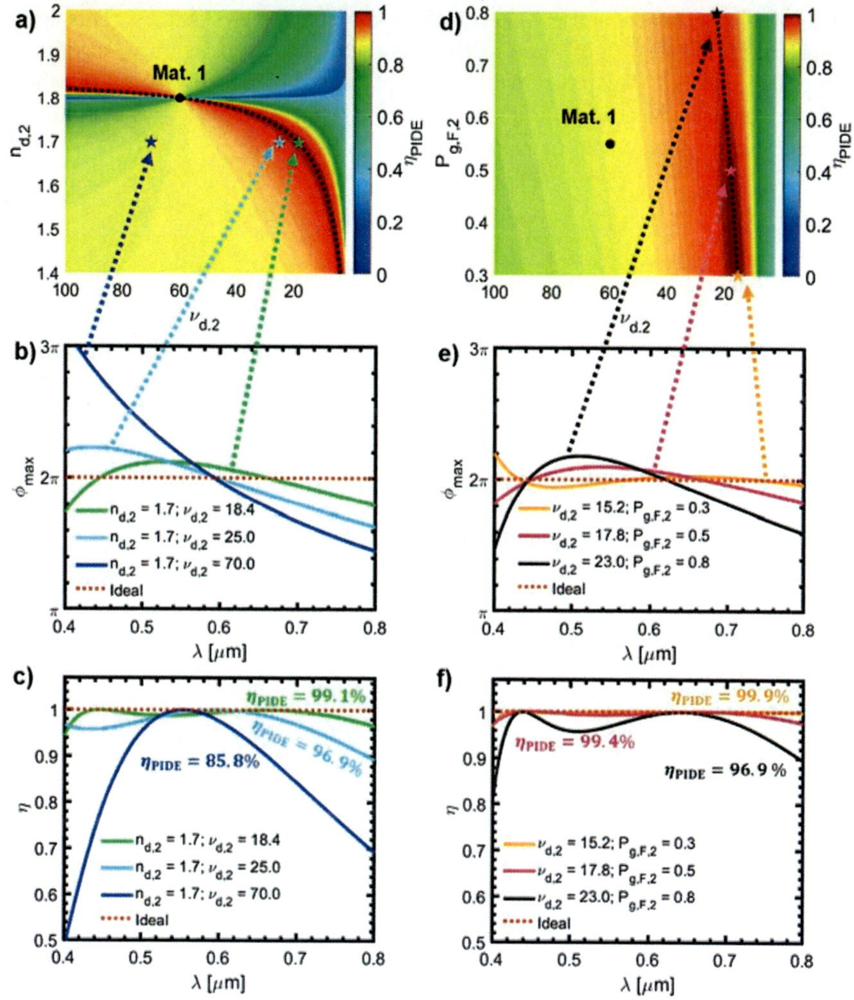

Fig. 5.11 **a** Polychromatic integral diffraction efficiency from the thin element approximation (η_{PIDE}^{TEA}) over the refractive index at the d-line and the Abbe number of $n_2(\lambda)$ ($n_{d,2}$ and $\nu_{d,2}$). The parameters that describe $n_1(\lambda)$ were fixed at $n_{d,1} = 1.80$, $\nu_{d,1} = 60$, and $P_{g,F,1} = 0.55$ (black dot). The partial dispersion of $n_2(\lambda)$ was fixed at $P_{g,F,2} = 0.55$. The dashed black line visualizes the curve on which η_{PIDE}^{TEA} is maximized. **b** Phase profiles and **c** efficiency curves for different sets of $n_{d,2}$ and $\nu_{d,2}$ values. The corresponding values are highlighted by the stars in (**a**). **d**, **e**, **f** Analogous plots over $P_{g,F,2}$ and $\nu_{d,2}$ for $n_{d,2} = 1.70$. Reprinted with permission from [3] ©The Optical Society

dispersions at $P_{g,F,2} = P_{g,F,1} = 0.55$. I chose this value because it is an intermediate value, which lies in close vicinity to the $P_{g,F}$-values of all conventional optical materials (Fig. 4.1b). With these parameters fixed, I then successively varied $n_{d,2}$ and $\nu_{d,2}$ and, for each parameter combination, optimized the DOE's height (see (2.18)) for a maximum $\eta_{\mathrm{PIDE}}^{\mathrm{TEA}}$. Accordingly, Fig. 5.11a depicts $\eta_{\mathrm{PIDE}}^{\mathrm{TEA}}$ over $n_{\mathrm{d},2}$ and $\nu_{\mathrm{d},2}$. This figure demonstrates that a $\eta_{\mathrm{PIDE}}^{\mathrm{TEA}}$ of close to 100% is only reached in a narrow corridor of $n_{\mathrm{d},2}$ and $\nu_{\mathrm{d},2}$ values. This corridor is highlighted by the dashed black line, which represents a fit to the combinations of $n_{\mathrm{d},2}$ and $\nu_{\mathrm{d},2}$ for which $\eta_{\mathrm{PIDE}}^{\mathrm{TEA}}$ is maximized. This line can be phenomenologically described as follows:

$$n_{\mathrm{d},2}(\nu_{\mathrm{d},2}) = a\left[1 - \exp\left(-b(\nu_{\mathrm{d},2} - c)^{\frac{1}{3}}\right)\right], \tag{5.11}$$

where, for this specific choice of parameters ($n_{\mathrm{d},1} = 1.8$ $\nu_{\mathrm{d},1} = 60$, and $P_{\mathrm{g,F},1} = P_{\mathrm{g,F},2} = 0.55$), the parameters a, b and c take values of $a = 1.839$,$b = 0.9918$, and $d = 0.8782$. Below I generalize this finding by determining a general expression for the coefficients a, b and c in (5.11) for different values of $n_{\mathrm{d},1}$ and $\nu_{\mathrm{d},1}$.

To directly visualize the dependence of the DOEs' phase profiles on $\nu_{\mathrm{d},2}$, which quantifies the overall amount of dispersion of $n_2(\lambda)$, Fig. 5.11b depicts $\phi_{\max}(\lambda)$ for three different values of $\nu_{\mathrm{d},2}$ with $n_{\mathrm{d},2}$ fixed at $n_{\mathrm{d},2} = 1.7$. To additionally visualize the optimal wavelength independent behavior, the dashed red line in this figure depicts the function $\phi_{\max}(\lambda) = 2\pi$. The different curves in Fig. 5.11b show that $\phi_{\max}(\lambda)$ depends drastically on $\nu_{\mathrm{d},1}$: First, the green line in Fig. 5.11b highlights that if $n_{\mathrm{d},2}$ and $\nu_{\mathrm{d},2}$ are located on the dashed black line (Fig. 5.11a), $\phi_{\max}(\lambda)$ fulfills the condition $\phi_{\max}(\lambda) = 2\pi$ for two distinct wavelengths (λ_1 and λ_2). This demonstrates that the dashed black line in Fig. 5.11a describes the curve on which the DOEs are achromatic to the first degree, in the sense that, they fulfill the condition $\eta^{\mathrm{TEA}}(\lambda) = 100\%$ for two different wavelengths (Fig. 5.11c). Second, the cyan curve in Fig. 5.11b illustrates that if $\nu_{\mathrm{d},2}$ is moved away from the dashed black line (cyan star with $\nu_{\mathrm{d},2} = 9.8$ in Fig. 5.11a), the second wavelength (λ_2) is shifted out of the visible spectrum. Lastly, the blue curve in Fig. 5.11b shows that far away from the dashed black line, the slope of $\phi_{\max}(\lambda)$ and hence the overall phase difference across the visible spectrum are drastically increased. Accordingly, Fig. 5.11c, shows that this leads to a marked drop of the efficiency towards the edges of the spectrum. In conclusion, this analysis demonstrates that a suitable material combination consists of a first material ($n_1(\lambda)$) that is characterized by higher values of both $n_{\mathrm{d},1}$ and $\nu_{\mathrm{d},1}$ than the second material ($n_2(\lambda)$). To achieve efficiency achromatization to the first degree, $n_{\mathrm{d},1}/\nu_{\mathrm{d},1}$ and $n_{\mathrm{d},2}/\nu_{\mathrm{d},2}$ must then be precisely tailored according to (5.10). This shows that dispersion-engineering is the key to achieving efficiencies of close to 100% across a broad spectral range. In fact, below I demonstrate the generality of this finding for widely different examples and analyze the influence of the other parameters ($n_{\mathrm{d},1}$, $\nu_{\mathrm{d},1}$, $P_{\mathrm{g,F},1}$, and $P_{\mathrm{g,F},2}$).

As the second step, I now fix $n_{\mathrm{d},2}$ at $n_{\mathrm{d},2} = 1.7$ to investigate the influence of the dispersion of $n_2(\lambda)$ ($\nu_{\mathrm{d},2}$ and $P_{\mathrm{g,F},2}$) independently. To this end, Fig. 5.11d, depicts $\eta_{\mathrm{PIDE}}^{\mathrm{TEA}}$, as a function of these parameters. In this figure, the dashed black line again visualizes the parameter combinations for which $\eta_{\mathrm{PIDE}}^{\mathrm{TEA}}$ is maximized. It is evident that there is indeed only a narrow corridor of $\nu_{\mathrm{d},2}$ values in which efficiency achromatization can be achieved. On the other hand, the influence of $P_{\mathrm{g,F},2}$ on $\eta_{\mathrm{PIDE}}^{\mathrm{TEA}}$ is much less pronounced and high efficiencies can be obtained across a much wider range of

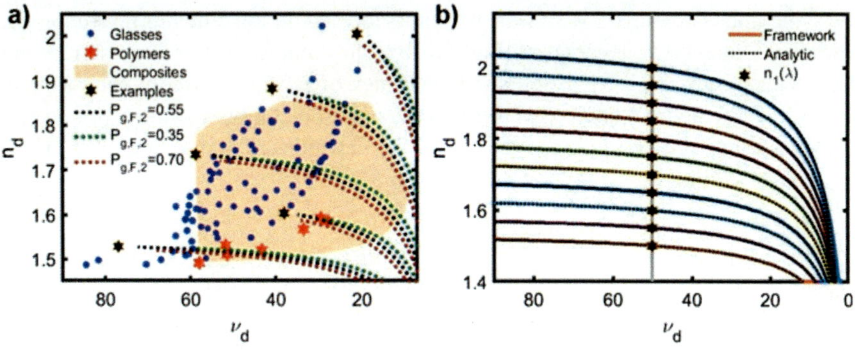

Fig. 5.12 **a** Lines in the Abbe diagram along which $\eta_{\text{PIDE}}^{\text{TEA}}$ is maximized (in analogy to Fig. 5.11 and (5.11)) for a wide range of different starting materials and different values of $P_{g,F,2}$. As the starting materials, the glasses N-PK51, N-LASF31A, F5, and P-SF68 (from left to right; see [44]), and a diamond in PMMA nanocomposite ($f = 30\%$; see Fig. 4.1) were chosen. To achieve efficiency achromatization, it is necessary to combine a high n_d/ν_d material with a second material with low values of n_d/ν_d. Varying $P_{g,F,2}$ merely leads to minor shift of the line on which efficiency achromatization is achieved. **b** Procedure used for determining the analytic design expressions. In analogy to the dotted black line in Fig. 5.11a, the colored lines were obtained from the design framework (see Sect. 5.3.1) for the different starting points ($n_1(\lambda)$) that are highlighted by the black stars. The dotted black lines were obtained from the analytic design expressions given in (5.11) and (5.12). The results from these expressions almost perfectly agree with the data obtained from the design framework. Reprinted with permission from [3] ©The Optical Society

$P_{g,F,2}$ values. However, decreasing $P_{g,F,2}$ along the dashed black line in Fig. 5.11d continuously increases $\eta_{\text{PIDE}}^{\text{TEA}}$. Figure 5.11e visualizes that this is because decreasing $P_{g,F,2}$ along this line gradually flattens $\phi_{\max}(\lambda)$. In fact, the orange curve in Fig. 5.11e demonstrates that, for low values of $P_{g,F,2}$, even a third wavelength (λ_3) emerges at which the condition $\phi_{\max}(\lambda) = 2\pi$ is fulfilled. Accordingly, Fig. 5.11f, shows that this allows for a much higher degree of efficiency achromatization ($\eta_{\text{PIDE}}^{\text{TEA}} \geq 99.9\%$). To maximize $\eta_{\text{PIDE}}^{\text{TEA}}$, it is therefore essential that the low n_d/ν_d material also has a low partial dispersion ($P_{g,F}$) and the high n_d/ν_d a high value of $P_{g,F}$ (see below).

To show that the findings from the previous paragraphs indeed hold true across the full range of n_d, ν_d, and $P_{g,F}$ values that are accessible with realistic transparent materials, Fig. 5.12 depicts the lines in the Abbe diagram on which $\eta_{\text{PIDE}}^{\text{TEA}}$ is maximized for five different starting points ($n_1(\lambda)$). Furthermore, for each of these materials, the lines for three different values of $P_{g,F,2}$ are included. These different lines show that changing $P_{g,F,2}$ only leads to a relatively minor shift of the corridor in which efficiency achromatization can be achieved. Therefore, it is a good first approximation to neglect the influence of $P_{g,F}$. This simplification now allows me to use my phenomenological relationship for $n_{d,2}(\nu_{d,2})$ (5.11) to determine an analytic design equation for broadband DOEs. To this end, I determined the dependence of the variables a, b, and c in (5.11) on $n_1(\lambda)$. To do so, I chose eleven different starting points ($n_1(\lambda)$) within a range of $1.5 \leq n_{d,1} \leq 2.0$ and a fixed value of $\nu_{d,1} = 50$ (see Fig. 5.12b). This restriction to a fixed value of $\nu_{d,1}$ does not limit the generality because the same line is obtained for all starting points located on this line. Furthermore, the lines for different $n_{d,1}$ never cross (Fig. 5.12b). The complete set of lines is therefore completely determined by $n_{d,1}$. For other values of $\nu_{d,1}$, it is simply necessary to vary $n_{d,1}$ until the line that passes through the required combination of $n_{d,1}$ and $\nu_{d,1}$ is found. Using this procedure, I found that the dependence of a, b, and c (5.11) on $n_{d,1}$ can be written as:

$$a(n'_{d,1}) = 1.071 n'_{d,1} - 0.07825, \qquad b(n'_{d,1}) = \frac{0.9555}{(n'_{d,1})^4} - 0.07825, \\ \text{and} \quad c(n'_{d,1}) = 53.59\sqrt{n'_{d,1}} - 16.59 n'_{d,1} - 41.17. \tag{5.12}$$

Figure 5.12b visualizes that the lines obtained from these phenomenological expressions ((5.11) and (5.12)) agree almost perfectly with the data obtained directly from my design framework (in analogy to Fig. 5.11). In addition, I verified these expressions with ten validation points that I didn't include in the fitting procedure. Note that I used the prime symbol on $n'_{d,1}$ to indicate that, for values other than $\nu_{d,1} = 50$, this parameter has to be adjusted until the line that passes through the required point is found. This shows that substituting the coefficients from (5.12) into (5.11) allows for directly identifying suitable values for $n_{d,2}$ and $\nu_{d,2}$ if $n_1(\lambda)$ is fixed. As shown previously, only small adjustments are required if the partial dispersions ($P_{g,F,1}$ or $P_{g,F,2}$) deviate significantly from 0.55.

Finally, the findings from this section directly lead me to my first two design rules for broadband EGs: (1) $n_1(\lambda)$ must be characterized by high values of $n_{d,1}$ and $\nu_{d,1}$, whereas $n_2(\lambda)$ must have low values of $n_{d,2}$ and $\nu_{d,2}$. To achieve first degree efficiency achromatization ($\eta_{PIDE}^{TEA} \approx 99.0\%$), these parameters ($n_{d,1}/\nu_{d,1}$ and $n_{d,2}/\nu_{d,2}$) must then be precisely tailored according to (5.11) and (5.12). And (2) to achieve higher degree efficiency achromatization ($\eta_{PIDE} \geq 99.9\%$), the first refractive index profile ($n_1(\lambda)$) must be characterized by a high $P_{g,F,1}$ and $n_2(\lambda)$ must have a low $P_{g,F,2}$.

Maximizing the range of AOIs and grating periods for which high efficiencies are maintained

As discussed previously, achieving high average efficiencies within the TEA throughout the visible spectrum (a high η_{PIDE}^{TEA}) is merely the first essential requirement for the widespread integration of DOEs into broadband systems. The second requirement is that a high η_{PIDE} must also be preserved across the full range of grating periods and AOIs that is required for such systems. Below, I systematically show using FEM simulations that this can be achieved by minimizing the EGs' heights (h) to reduce the amount of shadowing. Therefore, there are two distinct optimization targets for h. First, h must be optimized for a maximum η_{PIDE}^{TEA}. Second, h must be minimized to reduce the amount of shadowing. To determine how the latter requirement can be integrated into my design rules, Fig. 5.13a depicts h over $n_{d,2}$ as well as $\nu_{d,2}$. For this figure, all other quantities remained fixed as for Fig. 5.11a. Figure 5.13a shows that minimizing h can be achieved by maximizing $\Delta n_d = n_{d,1} - n_{d,2}$. However, it is of course critical to do so by choosing points on the dashed black line (from Fig. 5.11a)

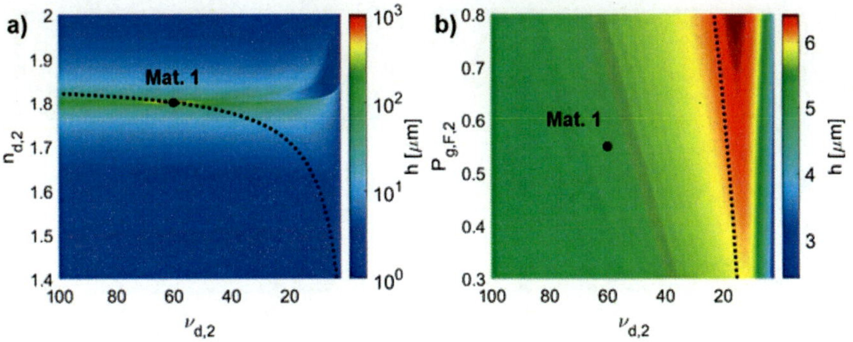

Fig. 5.13 EG-height (h) over **a** $n_{d,2}$ as well as $\nu_{d,2}$ and **b** $P_{g,F,2}$ as well as $\nu_{d,2}$. All other material parameters were fixed as for Fig. 5.11. The dashed black lines in both diagrams again indicate the curves on which the efficiencies are maximized (see Fig. 5.11a, d). Reprinted with permission from [3] ©The Optical Society

to achieve efficiency achromatization. The finding that Δn_d, and consequently the overall magnitude of the refractive index difference between the two layers, must be maximized can be readily understood from (2.18). This is because this equation demonstrates that h is inversely proportional to the magnitude of $\Delta n(\lambda)$. Finally, Fig. 5.13b illustrates that h depends only weakly on $\nu_{d,2}$ and $P_{g,F,2}$. These relatively small changes of h are a consequence of the optimization for a maximum η^{TEA}_{PIDE}, which leads to slightly different design wavelengths for different $\nu_{d,2}$ and $P_{g,F,2}$. This now leads me to my last design rule: (3) Δn_d must be maximized along the dashed black line on which η^{TEA}_{PIDE} is maximized. Accordingly, $\Delta \nu_d$ must also be maximized but, at the same time, precisely tailored according to (5.11) and (5.12).

Ideal design prototype with full parameter variation

To show that my design rules remain valid if all parameters are varied simultaneously, I performed an additional optimization. For this ideal design prototype, I optimized all six parameters ($n_{d,i}$, $\nu_{d,i}$, and $P_{g,F,i}$ for $i \in \{1, 2\}$) for a maximum η^{TEA}_{PIDE} and a minimal height (h). I included the full range of the axes of Fig. 5.11 in this optimization. Note that this includes regions which are far outside the regimes that are accessible with realistic materials (Fig. 4.1). Performing this optimization lead to: $n_{d,1} = 2.0$, $\nu_{d,1} = 78.7$, $P_{g,F,1} = 0.80$, as well as $n_{d,1} = 1.40$, $\nu_{d,2} = 2.0$, and $P_{g,F,2} = 0.30$. These parameters directly confirm my design rules. This is because, first, they show that $n_1(\lambda)$ is characterized by high values of $n_{d,1}$, $\nu_{d,1}$, and $P_{g,F,1}$, whereas $n_2(\lambda)$ is characterized by low values of all these parameters. Moreover, Δn_d is maximized completely within the permitted range, whereas $\Delta \nu_d$ is only maximized such that $n_{d,2}(\nu_{d,2})$ lies on the dashed black line on which η^{TEA}_{PIDE} is maximized (see Fig. 5.11a and (5.11)). Lastly, $P_{g,F,1}$ is fully maximized and $P_{g,F,2}$ completely minimized. To visualize that the efficiency profile of the optimized EG is almost perfectly achromatic, Fig. 5.14a presents its efficiency (η^{TEA}) over the wavelength. It is evident that this EG's η^{TEA} remains above 99.5% across the entire visible spec-

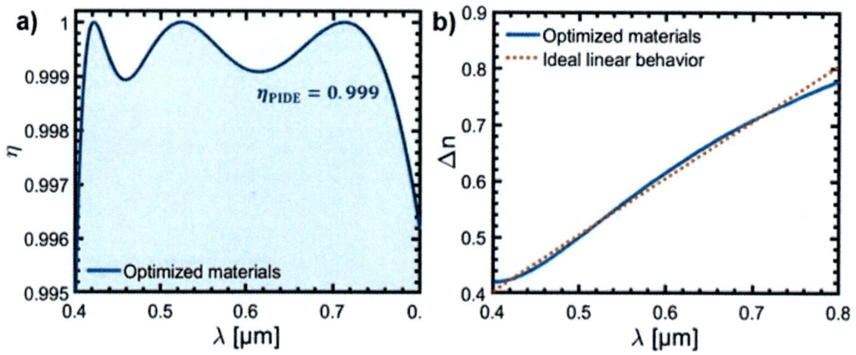

Fig. 5.14 **a** Efficiency from the TEA (η^{TEA}) over the wavelength (λ) for an ideal EG for which both refractive index profiles $n_1(\lambda)$ and $n_2(\lambda)$ were freely optimized for a maximum $\eta_{\mathrm{PIDE}}^{\mathrm{TEA}}$ and minimum height (h). The optimal parameters are: $n_{\mathrm{d},1} = 2.0$, $\nu_{\mathrm{d},1} = 78.7$, $P_{\mathrm{g,F},1} = 0.80$, as well as $n_{\mathrm{d},1} = 1.40$, $\nu_{\mathrm{d},2} = 2.0$, and $P_{\mathrm{g,F},2} = 0.30$. The optimal height is $h = 0.99$ μm. **b** $\Delta n(\lambda) = n_1(\lambda) - n_2(\lambda)$ of the optimized EG over the wavelength. It is evident that the optimized design is very close to the ideal linear relationship $\Delta n(\lambda) = \frac{\lambda}{h}$ (see (2.18)). Reprinted with permission from [3] ©The Optical Society

trum. Therefore, the EG has a $\eta_{\mathrm{PIDE}}^{\mathrm{TEA}}$ of 99.9%. In addition, Fig 5.14 illustrates that this was indeed achieved by tailoring the refractive index profile such that $\Delta n(\lambda)$ increases close to linearly with the wavelength (λ). Finally, this EG's height is only $h = 0.99$ μm because the refractive index difference Δn_{d} is unrealistically large.

5.3.2 Systematic Investigation of Nanocomposite-Enabled EGs

The framework and design rules that I developed in the previous sections are perfectly suited for the design of EGs. But, they are based on the TEA. Therefore, accurate numerical simulations are indispensable to quantify the EGs' actual efficiencies as a function of all relevant parameters. This is because shadowing effects (see Sect. 2.4) can only be fully accounted for through such simulations [46]. To quantify what performance can be expected depending on the material availability, I consequently now again use the FEM software JCMsuite and the procedure introduced in Sect. 5.1. For this purpose, I first use my design rules to systematically expand the two specific nanocomposite-enabled EGs from Table 5.1 by a wider range of prototype EGs. To be able to perform a systematic analysis, I do so by successively restricting the choice of filler materials. However, I always focus on the n_{d}, ν_{d} and $P_{\mathrm{g,F}}$ values of all materials to generalize my findings.

The first prototype system I use for investigating the influence of the material availability is the ideal nanocomposite-enabled EG (EG 2). As specified in Table 5.1, this EG is made of a ITO and a PMMA nanocomposite. From my design rules,

Table 5.2 Specifications of the additional nanocomposite-enabled (NE) EGs (EG 4 and EG 5) and the "common depth" (CD) EG 6, which is made of conventional polymers. Materials 1 and 2 correspond to the materials in the EGs' first and second layer, respectively. The percentages denote the respective volume fractions (f), h the EGs' heights and $\eta_{\text{PIDE}}^{\text{TEA}}$ the theoretical limit for the polychromatic integral diffraction efficiency that is obtained from the thin element approximation (TEA)

Name	Type	Material 1	Material 2	h (μm)	$\eta_{\text{PIDE}}^{\text{TEA}}$
EG 4	NE	35.0% dia. in PMMA	14.3% TiO_2 in PMMA	7.0	98.53%
EG 5	NE	35.0% ZrO_2 in PMMA	8.7% TiO_2 in PC	10.7	98.40%
EG 6	CD	PS	PC	86.6	96.80%

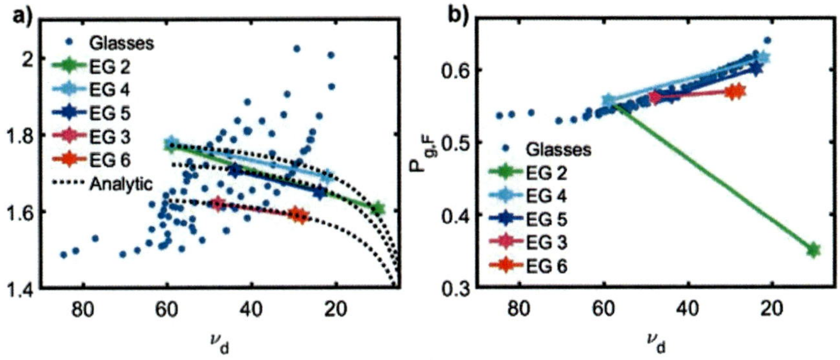

Fig. 5.15 Locations of all prototype EGs' constituent materials in the Abbe **a** and partial dispersion **b** diagrams (EG 2—EG 6 from Tables 5.1 and 5.2). In the legend, the different EGs are sorted by ascending heights. The range of materials included in the optimization of the EGs was systematically restricted to investigate the EGs' performance as a function of the material availability. The dashed black lines in (**a**) visualize that the locations of all material combinations agree well with the analytic design expressions given in (5.11) and (5.12). Reprinted with permission from [3] ©The Optical Society

it can now be readily understood why this device is distinguished by such a high performance. This is because the diamond in PMMA nanocomposite in its first layer (see Table 5.1) is distinguished by a high Abbe number (ν_d) and refractive index at the d-line (n_d; see Fig. 4.1a). Moreover, this material also has a relatively high partial dispersion ($P_{g,F}$; see Fig. 4.1b). In contrast, the ITO in PMMA nanocomposite in the EG's second layer is characterized by low values of ν_d and $P_{g,F}$ and a relatively small value of n_d. Therefore, this material combination fulfills all design rules to a very high degree and the EG hence exhibits a very high performance. This can also be directly seen from Fig. 5.15. In fact, the corresponding dashed black line in Fig. 5.15a shows that the material combination of this ideal nanocomposite-enabled EG (EG 2) agrees well with the analytic design expressions given in (5.11) and (5.12). The slight deviation of the low n_d/ν_d material from the dashed black line can

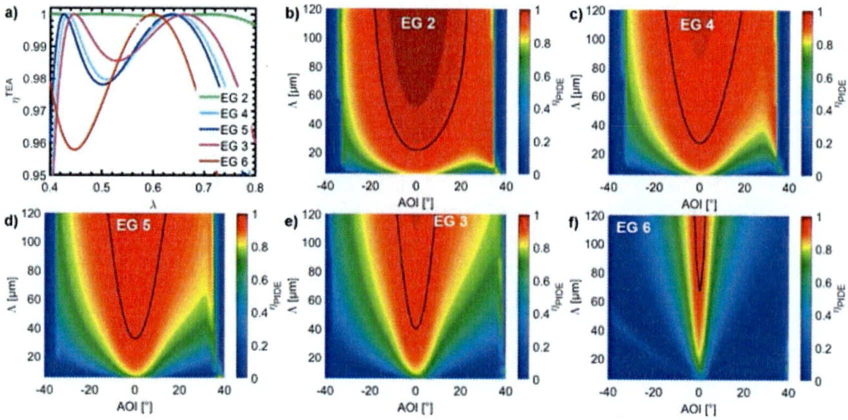

Fig. 5.16 **a** Efficiency as obtained from the TEA ((2.16) and (2.17)) over the wavelength for the different EGs listed in Tables 5.1 and 5.2. **b–f** Average efficiency in the visible spectrum (η_{PIDE}) in TM polarization as a function of the grating period (Λ) and the angle of incidence (AOI) as obtained from FEM simulations (JCMsuite) for EG 2—EG 6. In all figures, the dashed black lines highlight the regions in which η_{PIDE} remains within 3% of its maximum value. For each EG, this maximum value is obtained for $\Lambda = 120$ µm and AOI $= 0°$. The diffraction angle for the first order ($\theta_{q=1}$) can be readily calculated for all points in the diagrams using the grating equation (see (2.15)). Reprinted with permission from [3] ©The Optical Society

be understood from the fact that I derived the analytic design expressions for fixed partial dispersions of $P_{\mathrm{g,F},1} = P_{\mathrm{g,F},2} = 0.55$

To obtain a wider range of prototype EGs, I can now directly use my design rules. Specifically, from the Abbe diagram in Fig. 4.1 it is evident that ZrO_2 would be a less favorable alternative to diamond as the filler material for the high $n_\mathrm{d}/\nu_\mathrm{d}$ layer, whereas a TiO_2 nanocomposite or even a conventional polymer could be used in the second layer. Using these materials, I designed three additional EGs (EG 4—EG 6 in Table 5.2). For all these prototype EGs, Fig. 5.15 presents the constituent materials' locations in the Abbe (Fig. 5.15a) and partial dispersion diagrams (Fig. 5.15b). In agreement with my design rules, all EGs are comprised of a high $n_\mathrm{d}/\nu_\mathrm{d}$ material and a low $n_\mathrm{d}/\nu_\mathrm{d}$ material. This is also visualized by the dashed black lines, which I obtained from the analytic design expressions given in (5.11) and (5.12). However, Fig. 5.15b shows that only the ideal EG's (EG 2) low $n_\mathrm{d}/\nu_\mathrm{d}$ material also exhibits a low $P_{\mathrm{g,F}}$. This explains why only this device is distinguished by a $\eta_{\mathrm{PIDE}}^{\mathrm{TEA}}$ of 99.9%. For all other EGs, the low $n_\mathrm{d}/\nu_\mathrm{d}$ material has a partial dispersion of $P_{\mathrm{g,F}} \geq 0.55$. Therefore, these devices only exhibit two efficiency peaks (see Fig. 5.16) and hence a $\eta_{\mathrm{PIDE}}^{\mathrm{TEA}}$ below 99.0%.

Real-world performance for all prototype systems

To systematically investigate the performance of all prototype EGs, I again, as discussed in Sect. 5.1, used FEM simulations to quantify the EGs' η_{PIDE} as a function of the AOI and the grating period. Accordingly, Figs. 5.16b–f depict the EGs' η_{PIDE} as a function of these parameters. In this figure, I included the data for EG 2 and

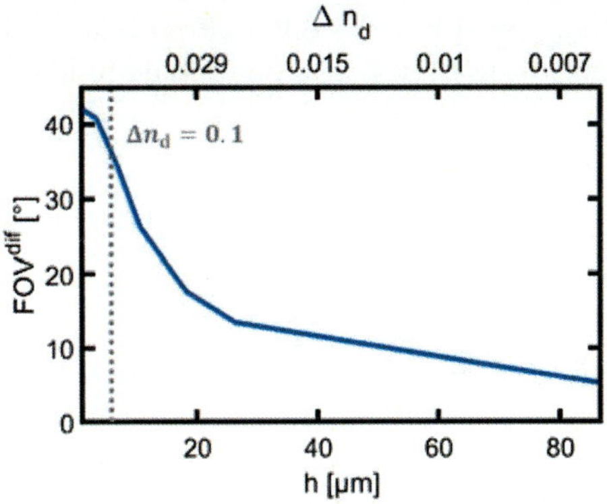

Fig. 5.17 Full field of view (FOV^{dif}) across which an EG maintains a η_{PIDE} that remains within 3% of its maximum value as a function of the EG's height (h) at a grating period of $\Lambda = 120\ \mu m$. The dashed grey line visualizes that, if the refractive index difference exceeds $\Delta n_d = 0.1$, the FOV^{dif} decreases quickly. Reprinted with permission from [3] ©The Optical Society

EG 3, which were already presented in Fig 5.4, to directly visualize the impact of a gradual restriction of the choice of filler materials.

The dashed black lines in all figures again visualize the region within which η_{PIDE} remains within 3% of its maximum. These lines directly demonstrate that the region of AOIs and grating periods for which a high η_{PIDE} is maintained decreases quickly in size with increasing heights. To directly visualize this relationship, Fig. 5.17 depicts the EGs' full field of view (FOV^{dif}) as a function of the EGs' heights. Here I define FOV^{dif} as the full range of AOIs for which η_{PIDE} remains within 3% of the maximum value that is achieved within the entire parameter range included in the simulation. For each EG, this maximum is obtained for perpendicular incidence ($\theta_{AOI} = 0°$) and the largest grating period of $\Lambda = 120\ \mu m$. In this figure, I also included the respective data for the single-layer EG (EG 1 in Table 5.1; see Fig. 5.1b) and the state-of-the-art common depth EG (EG S2 in Table A.1; see Fig. A.11b in the appendix) at heights of $h_{EG1} = 1.1\ \mu m$ and $h_{EGS2} = 26.3\ \mu m$, respectively. All these data combined (Fig. 5.17) show that the EGs' FOV (FOV^{dif}) decreases quickly with increasing heights even for $h < 10\ \mu m$. In fact, the grey line in Fig. 5.17 demonstrates that the onset of the drop in FOV^{dif} occurs around $\Delta n_d = 0.1$. Note that, as discussed in Sect. 5.1, I assumed the EGs' materials to extend to infinity for all simulations. As discussed in Sect. 5.2.4, the EGs' FOV^{dif} will be significantly larger, if the devices are placed into air as the surrounding medium. This is because the incident light is refracted at the additional interfaces according to Snell's law. For example, if I assume a refractive index of $n = 1.6$ for the base layer, the value of $FOV^{dif} = 37°$, which is achieved at the threshold of $\Delta n_d = 0.1$, corresponds to a value of $FOV^{dif} \approx 61°$ (see Fig. 5.9). In conclusion, this shows that the range of AOIs that is incident on a DOE directly determines the required value of Δn_d. In fact, as already discussed, to maximize FOV^{dif} and hence design EGs that cover the requirements of a wide range of optical systems (see sec. 5.2.4), refractive index differences of $\Delta n_d \geq 0.1$ are required. For the nanocomposite-enabled EGs, this condition is only fulfilled by EG 2.

5.3.3 Material Combinations for High Performance EGs

In Sect. 5.2, I have already shown that the ideal nanocomposite-enabled device (EG 2), fulfills all requirements of most broadband imaging systems. Now I can use my design rules together with the requirement for $\Delta n_d \geq 0.1$ to generalize this finding and investigate what material properties are required for the design of such high-performance EGs. To systematically do so, I considered the full range of materials depicted in the Abbe diagram in Fig. 5.18a (glasses, polymers and nanocomposites). Furthermore, since I have already shown that the influence of the partial dispersion is relatively small, I fixed the partial dispersion of the second material at $P_{g,F,2} = 0.55$. Then I chose a wide range of different materials for the first layer along the high n_d/ν_d edge in the Abbe diagram (Fig. 5.18a). For each of these materials, I subsequently used my framework to determine the lines along which $\eta_{\text{PIDE}}^{\text{TEA}}$ is maximized (5.11). Finally, I evaluated within what range on these lines the requirement for $\Delta n_d \geq 0.1$ can be fulfilled. The blue and green areas in the Abbe diagram in Fig. 5.18 visualize the results of this procedure. Specifically, the blue area denotes the region within which the high n_d/ν_d material must be located, whereas the green area highlights the corresponding region for the low n_d/ν_d material. I emphasize, however, that efficiency achromatization is possible by combining any two points along these lines, but a refractive index difference of $\Delta n_d \geq 0.1$ can only be achieved within the colored regions.

The shaded regions in Fig. 5.18 demonstrate that high-performance EGs can be designed using a wide range of readily available materials as the high n_d/ν_d material. For example, ZrO_2 nanocomposites or glasses are suitable choices for this material. However, materials that can be used as the low n_d/ν_d material are not yet commercially available. In fact, to achieve a value of $\eta_{\text{PIDE}}^{\text{TEA}} \geq 99.9\%$, this material should also have a low partial dispersion ($P_{g,F,2}$). As discussed previously, ITO nanocomposites offer the possibility to access this region. However, future experimental research is required to evaluate whether these materials not only fulfill the optical requirements but can also allow for mass production. Since a certain refractive index profile can generally be achieved with different chemical compositions, it could also be possible to develop other materials that achieve the same properties. In this context, the fact that the heights of high-performance EGs are below 10 μm is a key factor. This is because conventional optical materials must generally maintain a high transmittance for lenses with thicknesses exceeding several centimeters. Therefore, the absorption coefficients of conventional optical materials must be almost zero. However, according to Beer's law, the transmitted intensity decreases exponentially with the propagation distance. Therefore, for thin layers, a much higher absorption coefficient still leads to the same transmittance. There is consequently much more freedom in the design of optical materials that are only used as thin layers. Specifically, the

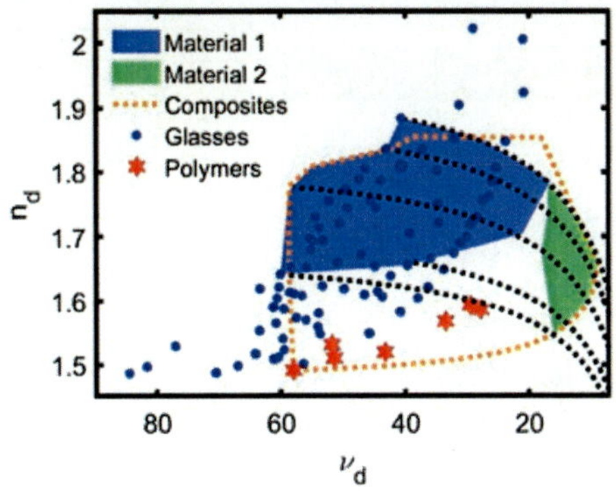

Fig. 5.18 Regions in the Abbe diagram within which materials that allow for the design of high-performance EGs must be located. The blue area encompasses region in which the first material (high n_d/ν_d material) must be located, whereas the green area visualizes the corresponding region for the second material (low n_d/ν_d material). The dotted black lines represent the curves on which $\eta_{\mathrm{PIDE}}^{\mathrm{TEA}}$ is maximized for different starting points along the high n_d/ν_d edge of the region that is covered by materials in Fig. 4.1a (glasses, polymers, and nanocomposites). The partial dispersion of the second material was fixed at $P_{g,F,2} = 0.55$. The remaining (inner) edges of both the green and the blue areas follow from the requirement for $\Delta n_d \geq 0.1$. Reprinted with permission from [3] ©The Optical Society

materials' dispersion can be increased by adding additional optical resonances to the system. As discussed in Sect. 2.1, this is intrinsically linked to increasing the amount of absorption. For thin layers in the order of 10 μm, it is consequently possible to add more optical resonances to the system while still maintaining a high transmittance. This can enable the design of materials that are located far outside the dispersion regimes of conventional optical materials (Fig. 4.1).

5.4 Towards Broadband Metalenses and GRIN DOEs

Until now, I have focused on échelette-type gratings (EGs) because such DOEs can be readily mass produced using standard replication techniques [47]. However, as discussed in the introduction, several other technologies for DOEs have been developed. For example, one technology that has recently received a lot of interest are metagratings [27–33, 48–59]. Here I define metagratings as DOEs that are composed of distinct sub-wavelength scatterers. Since all technologies for DOEs generally suffer from strongly wavelength-dependent efficiencies, a key question consequently is whether my design formalism can also be generalized to other technologies like metagratings.

Before I address specific technologies, it is first essential to discuss what physical mechanisms are, in general, suitable for the design of DOEs that fulfill all requirements of broadband imaging systems. In practice, most DOEs rely on propagation

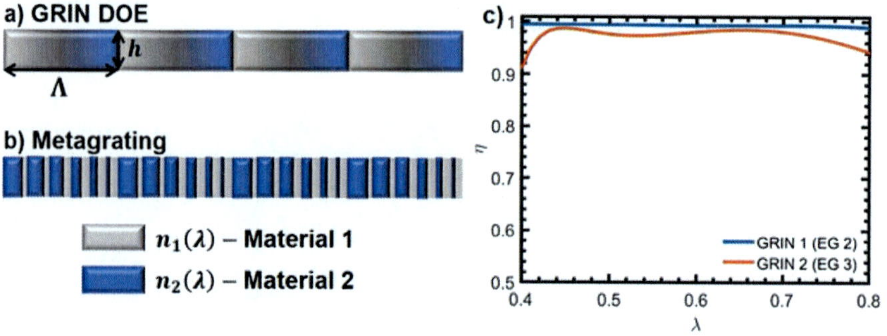

Fig. 5.19 Structure of **a** a GRIN DOE and **b** a metagrating. **c** Efficiency as obtained from FEM simulations (JCMsuite) over the wavelength for two GRIN DOEs, which are made of the same materials as EG 2 and EG 3 (see Table 5.1). For these simulations, a grating period of $\Lambda = 120\,\mu m$ and an AOI of 0° was used. Reprinted with permission from [3] ©The Optical Society

delays [12, 14, 21, 28, 50, 51]. Alternatively, resonant nanoparticles [14, 28, 48, 49, 60] or, if the incident light is circularly polarized, the so-called Pancharatnam-Berry phase (geometric phase) [14, 28, 29, 51, 54] can be used. However, around each optical resonance, the phase delay depends drastically on the wavelength. In fact, this is the fundamental mechanism that is utilized to implement the phase profiles of DOEs using resonant sub-wavelength particles. But, as evident from (2.17), this also leads to a drastic wavelength-dependence of the diffraction efficiency. Devices relying on this effect are consequently unsuited for broadband applications. Moreover, the geometric phase is intrinsically polarization-dependent and is thus not suited for systems that operate with unpolarized light. Therefore, the only mechanism that appears to be suited for conventional imaging systems is the propagation delay. In fact, this is also the mechanism that is utilized in EGs.

A wave that propagates through a material with a refractive index of $n(\lambda)$ accumulates a phase delay of $\phi(z, \lambda) = \frac{2\pi n(\lambda)}{\lambda} z$. This equation directly shows that propagation delays allow for implementing a DOE's phase profile by either varying the propagation distance (z) or the refractive index ($n(\lambda)$) as a function of the position. While EGs are based on the former approach, the latter leads to the concept of a gradient refractive index (GRIN) DOE (see illustration in Fig. 5.19). For a GRIN DOE, I can express the maximum phase delay per period ($\phi_{\max}(\lambda)^{\mathrm{TEA}}$) as the phase difference that arises from the refractive index difference between the two edges of each period. As visualized in Fig. 5.19a, these phase delays are determined solely by $n_1(\lambda)$ and $n_2(\lambda)$. Within the TEA, the maximum phase delay per period can consequently be written as $\phi_{\max}^{\mathrm{TEA}}(\lambda) = \frac{2\pi h}{\lambda}\left[n_1(\lambda) - n_2(\lambda)\right]$. Since this relationship is identical to the corresponding expression for EGs (see (2.17)), the concept of a GRIN DOE is mathematically equivalent to that of an EG within the TEA. This shows that my design rules can also be directly applied to GRIN DOEs. I disclosed this concept of a single-layer efficiency achromatized DOE in a recent patent application [61]. To show that high-performance GRIN DOEs can indeed be designed using the same material combinations as for EGs, I performed additional FEM simulations for two GRIN DOEs. For these simulations, I used the same materials as for EG 2 and EG 3 (see Table 5.1) and chose an AOI of 0° as well as a grating period of $\Lambda = 120\ \mu$m. The resulting efficiency curves are depicted in Fig. 5.19c. It is evident that the GRIN DOEs indeed also exhibit high efficiencies throughout the visible spectral range. Furthermore, the efficiency curves closely resemble those that were obtained for the corresponding EGs (Fig. 5.4a, c). Finally, I show in the appendix (Sect. A.7) that the dependence of a GRIN DOE's η_{PIDE} on the AOI and the grating period is essentially identical to the behavior of the corresponding EG. These results confirm that my design rules also enable the design of high-performance GRIN DOEs. Therefore, it is possible to integrate the functionality of a diffractive lens (DL) directly into the volume of a refractive lens. With the advent of novel fabrication technologies, in particular 3D printing, this approach could be used to combine the functionalities of both a DL and a refractive lens in the same physical element.

The refractive index profiles of the GRIN-type DOEs, which I discussed so far, are varied continously as a function of the position. However, it is also possible to sample the required refractive index profiles only in discrete increments by designing a grating that is composed of distinct sub-wavelength scatterers. This approach leads to the concept of a metagrating. For such devices, different regimes must be distinguished. First, if all dimensions and separations are deeply subwavelength, the ensemble of sub-wavelength building blocks simply acts as an effective medium. In analogy to three dimensional nanocomposites, the local refractive indices can consequently be readily adjusted by varying the spatial density of the scatterers [13, 14]. In this regime, my design rules for $n_1(\lambda)$ and $n_2(\lambda)$ hence directly apply to the effective refractive indices at the edges of each period. In contrast, if the distances between the building blocks reach a value of around $0.5 \cdot \lambda$, the coupling between the different building blocks is significantly reduced. In this regime, every building block can be approximately treated as a waveguide that has little mode overlap to the neighboring elements [11, 14]. This has already been demonstrated more than 20 years ago [11] and discussed in several more recent publications since then [14, 29, 50, 51]. In this waveguiding regime, the local phase delays are determined by the waveguides' mode indices. A DOE's phase profile can hence be realized by varying the waveguides' properties, that is, their geometries, dimensions, or materials as a function of the position. To apply my design rules to such metagratings, the mode indices at the edges of each period must consequently be described by their effective $n_{\mathrm{d,eff}}$, $\nu_{\mathrm{d,eff}}$, and $P_{\mathrm{g,F,eff}}$ values. The key to designing broadband metagratings and metalenses then is the design of waveguide geometries and compositions that allow for adjusting the waveguide dispersion such that my design rules are fulfilled. In the appendix (Sect. A.8), I analyze this aspect in more detail and suggest approaches for how this can be achieved. An advantage of metagratings compared to EGs is that the exploitation of the waveguiding regime can mitigate the impact of shadowing [12, 14, 27]. This is because, if the waveguides are operated in the single mode regime, the incident light can robustly couple to the fundamental mode and consequently always experiences the same phase delay. I emphasize, however, that introducing a discretization of the phase profile decreases the maximum achievable efficiencies

[20]. Therefore, it appears likely that designing metagratings with efficiencies of close to 100% across a broad spectral range will turn out to be an insurmountable task.

Finally, for metagratings, yet another regime is reached if the distances between the building blocks are increased beyond the waveguiding regime. At some point, the structure then starts to support multiple propagating Bloch modes and can therefore no longer be treated as a GRIN-type DOE. It has been demonstrated that optimizing a structure with complicated geometries in this "multiple Bloch modes" regime also allows for the design of gratings that achieve very high efficiencies [62]. However, the interference pattern of multiple Bloch modes is generally very sensitive to the different parameters, that is, the wavelength, the AOI, and the grating periods. These fundamental issues have been highlighted by Lalanne et al. in Ref [14]. Therefore, this approach is most likely not suitable for the design of DOEs that maintain a high performance across a wide range of wavelengths, AOIs, and grating periods.

5.4.1 Nanocomposites Allow for the Design of Diffractive Optical Elements that Fulfill all Requirements of High-End Broadband Optical Systems

In summary, in this chapter, I have shown that highly-efficient broadband diffractive optical elements can be designed through dispersion-engineering. In fact, this analysis has demonstrated that nanocomposites are an ideal material platform for the design of such DOEs. Furthermore, I have shown that such dispersion-engineered materials allow for the design of highly efficient diffractive lenses (DLs) that could finally unlock the potential of DLs for broadband optical systems. To build on these findings, I, in the following chapter, investigate if dispersion-engineering can also provide significant benefits for purely refractive optical elements and systems. Finally, I then analyze whether novel dispersion-engineered materials could push smartphone cameras beyond their current limits.

References

1. D. Werdehausen, S. Burger, I. Staude, T. Pertsch, M. Decker, Dispersion-engineered nanocomposites enable achromatic diffractive optical elements. Opt. **6**(8), 1031 (2019)
2. D. Werdehausen, S. Burger, I. Staude, T. Pertsch, M. Decker, Flat optics in high numerical aperture broadband imaging systems. J. Opt. **22**(6), 065607 (2020)
3. D. Werdehausen, S. Burger, I. Staude, T. Pertsch, M. Decker, General design formalism for highly efficient flat optics for broadband applications. Opt. Exp. **28**(5), 6452–6468 (2020)
4. T. Ogata, R. Yagi, N. Nakamura, Y. Kuwahara, S. Kurihara, Modulation of polymer refractive indices with diamond nanoparticles for metal-free multilayer film mirrors. ACS Appl. Mater. Interfaces **4**(8), 3769–72 (2012)

5. H. Liu, X. Zeng, X. Kong, S. Bian, J. Chen, A simple two-step method to fabricate highly transparent ITO/polymer nanocomposite films. Appl. Surface Sci. **258**(22), 8564–8569 (2012)
6. E. Ōsawa, Recent progress and perspectives in single-digit nanodiamond. Diamond Related Mater. **16**(12), 2018–2022 (2007)
7. Z. Chen, Pixelligent zirconia nano-crystals for OLED applications, in white paper (2014)
8. D. Russel, A. Stabell, Scaling-up pixelligent nanocrystal dispersions, in *White Paper* (2016)
9. Z. Chen, J. Wang, Pixelligent internal light extraction layer for OLED lighting, in *White Paper* (2014)
10. T. Stone, N. George, Hybrid diffractive-refractive lenses and achromats. Appl. Opt. **27**(14), 2960–71 (1988)
11. P. Lalanne, Waveguiding in blazed-binary diffractive elements. J. Opt. Soc. Am. A **16**(10), 2517 (1999)
12. P. Lalanne, S. Astilean, P. Chavel, E. Cambril, H. Launois, Blazed binary subwavelength gratings with efficiencies larger than those of conventional échelette gratings. Opt. Lett. **23**(14), 1081 (1998)
13. P. Lalanne, S. Astilean, P. Chavel, E. Cambril, H. Launois, Design and fabrication of blazed binary diffractive elements with sampling periods smaller than the structural cutoff. J. Opt. Soc. Am. A **16**(5), 1143–1156 (1999)
14. P. Lalanne, P. Chavel, Metalenses at visible wavelengths: past, present, perspectives. Laser Photon. Rev. **11**(3), 1600295 (2017)
15. P. Lalanne, J.P. Hugonin, P. Chavel, Optical properties of deep lamellar Gratings: a coupled Bloch-mode insight. J. Lightwave Technol. **24**(6), 2442–2449 (2006)
16. M.-S. L. Lee, P. Lalanne, J.-C. Rodier, E. Cambril, Wide-field-angle behavior of blazed-binary gratings in the resonance domain. Opt. Lett. **25**(23), 1690 (2000)
17. C. Ribot, M.-S.L. Lee, S. Collin, S. Bansropun, P. Plouhinec, D. Thenot, S. Cassette, B. Loiseaux, P. Lalanne, Broadband and efficient diffraction. Adv. Opt. Mater. **1**(7), 489–493 (2013)
18. C. Sauvan, P. Lalanne, M.-S.L. Lee, Broadband blazing with artificial dielectrics. Opt. Lett. **29**(14), 1593 (2004)
19. C.A. Palmer, E.G. Loewen, *Diffraction Grating Handbook* (Newport Corporation New York, 2005)
20. G.J. Swanson, *Binary Optics Technology: The Theory and Design of Multi-level Diffractive Optical Elements* (Report, Lincoln Laboratory Massachusetts Institute of Technology, 1989)
21. S. Banerji, M. Meem, A. Majumder, F.G. Vasquez, B. Sensale-Rodriguez, R. Menon, Imaging with flat optics: metalenses or diffractive lenses? Opt. **6**(6), 805 (2019)
22. G. Kim, J.A. Dominguez-Caballero, R. Menon, Design and analysis of multi-wavelength diffractive optics. Opt. Exp. **20**(3), 2814–23 (2012)
23. N. Mohammad, M. Meem, X. Wan, R. Menon, Full-color, large area, transmissive holograms enabled by multi-level diffractive optics. Sci. Rep. **7**(1), 5789 (2017)
24. P. Wang, N. Mohammad, R. Menon, Chromatic-aberration-corrected diffractive lenses for ultra-broadband focusing. Sci. Rep. **6**, 21545 (2016)
25. H. Gross, W. Singer, M. Totzeck, F. Blechinger, B. Achtner, *Handbook of Optical Systems*, vol. 1 (Wiley-VCH, Berlin, 2005)
26. A. Small, Spherical aberration, coma, and the Abbe sine condition for physicists who don't design lenses. Am. J. Phys. **86**(7), 487–494 (2018)
27. M. Decker, W.T. Chen, T. Nobis, A.Y. Zhu, M. Khorasaninejad, Z. Bharwani, F. Capasso, J. Petschulat, Imaging performance of polarization-insensitive metalenses. ACS Photon. (2019)
28. H. Liang, A. Martins, B.-H.V. Borges, J. Zhou, E.R. Martins, J. Li, T.F. Krauss, High performance metalenses: numerical aperture, aberrations, chromaticity, and trade-offs. Optica **6**(12), 1461–1470 (2019)
29. W.T. Chen, A.Y. Zhu, V. Sanjeev, M. Khorasaninejad, Z. Shi, E. Lee, F. Capasso, A broadband achromatic metalens for focusing and imaging in the visible. Nat. Nanotechnol. **13**(3), 220–226 (2018)

30. W.T. Chen, A.Y. Zhu, J. Sisler, Z. Bharwani, F. Capasso, A broadband achromatic polarization-insensitive metalens consisting of anisotropic nanostructures. Nat. Commun. **10**(1), 355 (2019)
31. W.T. Chen, A.Y. Zhu, J. Sisler, Y.W. Huang, K.M.A. Yousef, E. Lee, C.W. Qiu, F. Capasso, Broadband achromatic metasurface-refractive optics. Nano Lett. **18**(12), 7801–7808 (2018)
32. M. Khorasaninejad, Z. Shi, A.Y. Zhu, W.T. Chen, V. Sanjeev, A. Zaidi, F. Capasso, Achromatic Metalens over 60 nm bandwidth in the visible and metalens with reverse chromatic dispersion. Nano Lett. **17**(3), 1819–1824 (2017)
33. S. Wang et al., A broadband achromatic metalens in the visible. Nat. Nanotechnol. **13**(3), 227–232 (2018)
34. J.A. Davison, M.J. Simpson, History and development of the apodized diffractive intraocular lens. J. Cataract Refractive Surgery **32**(5), 849–58 (2006)
35. A.A. Kazemi, B. Kress, T. Starner, B.C. Kress, S. Thibault, A review of head-mounted displays (HMD) technologies and applications for consumer electronics, in *Proceedings SPIE 8720, Photonic Applications for Aerospace, Commercial, and Harsh Environments IV* (2013), p. 87200A
36. G.I. Greisukh, E.G. Ezhov, A.V. Kalashnikov, S.A. Stepanov, Diffractive-refractive correction units for plastic compact zoom lenses. Appl. Opt. **51**(20), 4597–604 (2012)
37. G.I. Greisukh, E.G. Ezhov, I.A. Levin, S.A. Stepanov, Design of achromatic and apochromatic plastic micro-objectives. Appl. Opt. **49**(23), 4379–84 (2010)
38. T. Nakai, H. Ogawa, Research on multi-layer diffractive optical elements and their application to camera lenses, in *Diffractive Optics and Micro-Optics*. Optical Society of America, 2002, DMA2
39. G.I. Greisukh, E.G. Ezhov, S.A. Stepanov, Diffractive-refractive hybrid corrector for achroand apochromatic corrections of optical systems. Appl. Opt. **45**(24), 6137 (2006)
40. J.M. Trapp, T.G. Jabbour, G. Kelch, T. Pertsch, M. Decker, Hybrid refractive holographic single vision spectacle lenses. J. Eur. Opt. Soc.-Rapid Publ. **15**(1), 14 (2019)
41. S. Schmidt, S. Thiele, A. Herkommer, A. Tunnermann, H. Gross, Rotationally symmetric formulation of the wave propagation method-application to the straylight analysis of diffractive lenses. Opt. Lett. **42**(8), 1612–1615 (2017)
42. R. Kingslake, *Lenses in Photography: The Practical Guide to Optics for Photographers* (Barnes, 1963)
43. N. Sultanova, S. Kasarova, I. Nikolov, Dispersion properties of optical polymers. Acta Physica Polonica-Ser. General Phys. **116**(4), 585 (2009)
44. Schott, Optical Glass 2020. Technical Report Schott AG (2020)
45. P. Hartmann, Optical glass: deviation of relative partial dispersion from the normal line-need for a common definition. Opt. Eng. **54**(10), 105112 (2015)
46. O. Sandfuchs, R. Brunner, D. Pätz, S. Sinzinger, J. Ruoff, Rigorous analysis of shadowing effects in blazed transmission gratings. Opt. Lett. **31**(24), 3638 (2006)
47. M.T. Gale, Replication techniques for diffractive optical elements. Microelectron. Eng. **34**(3–4), 321–339 (1997)
48. M. Decker, I. Staude, Resonant dielectric nanostructures: a low-loss platform for functional nanophotonics. J. Opt. **18**(10), 103001 (2016)
49. M. Decker, I. Staude, M. Falkner, J. Dominguez, D.N. Neshev, I. Brener, T. Pertsch, Y.S. Kivshar, High-efficiency dielectric Huygens' surfaces. Advanced Optical Materials **3**(6), 813–820 (2015)
50. S.J. Byrnes, A. Lenef, F. Aieta, F. Capasso, Designing large, high-efficiency, high-numerical-aperture, transmissive meta-lenses for visible light. Opt. Exp. **24**(5), 5110–5124 (2016)
51. P. Genevet, F. Capasso, F. Aieta, M. Khorasaninejad, R. Devlin, Recent advances in planar optics: from plasmonic to dielectric metasurfaces. Opt. **4**(1), 139 (2017)
52. B. Groever, W.T. Chen, F. Capasso, Meta-lens doublet in the visible region. Nano Lett. **17**(8), 4902–4907 (2017)
53. M. Khorasaninejad, F. Capasso, Metalenses: Versatile multifunctional photonic components. Sci. **358**(6367) (2017)

54. M. Khorasaninejad, W.T. Chen, R.C. Devlin, J. Oh, A.Y. Zhu, F. Capasso, Metalenses at visible wavelengths: diffraction-limited focusing and subwavelength resolution imaging. Science **352**(6290), 1190–4 (2016)
55. M. Khorasaninejad, A.Y. Zhu, C. Roques-Carmes, W.T. Chen, J. Oh, I. Mishra, R.C. Devlin, F. Capasso, Polarization-insensitive metalenses at visible wavelengths. Nano Lett. **16**(11), 7229–7234 (2016)
56. R. Sawant, P. Bhumkar, A.Y. Zhu, P. Ni, F. Capasso, P. Genevet, Mitigating chromatic dispersion with hybrid optical metasurfaces. Adv. Mater. **31**(3), e1805555 (2019)
57. A. She, S. Zhang, S. Shian, D.R. Clarke, F. Capasso, Adaptive metalenses with simultaneous electrical control of focal length, astigmatism, and shift. Sci. Adv. **4**(2), eaap9957 (2018)
58. N. Yu, P. Genevet, M.A. Kats, F. Aieta, J.P. Tetienne, F. Capasso, Z. Gaburro, Light propagation with phase discontinuities: generalized laws of reflection and refraction. Science **334**(6054), 333–7 (2011)
59. A.Y. Zhu, W.-T. Chen, M. Khorasaninejad, J. Oh, A. Zaidi, I. Mishra, R.C. Devlin, F. Capasso, Ultra-compact visible chiral spectrometer with meta-lenses. APL Photon. **2**(3), 036103 (2017)
60. A.I. Kuznetsov, A.E. Miroshnichenko, M.L. Brongersma, Y.S. Kivshar, B. Luk'yanchuk, Optically resonant dielectric nanostructures. Sci. **354**(6314), aag2472 (2016)
61. D. Werdehausen, M. Decker, Diffraktives optisches Element, Verfahren zum Entwerfen einer effizienzachromatisierten diffraktiven Struktur und Verfahren zur Herstellung eines effizienzachromatisierten diffraktiven Elementes. German Patent Application DE102019109944.7 (2019)
62. D. Sell, J. Yang, S. Doshay, R. Yang, J.A. Fan, Large-angle, multifunctional metagratings based on freeform multimode geometries. Nano Lett. **17**(6), 3752–3757 (2017)

Chapter 6
The Potential of Nanocomposites for Optical Design

Nanocomposites allow for controlling the magnitude and dispersion of the effective refractive index within broad regions (see Chap. 4). Exploiting this tunability offers a high potential for improving the performance or reducing the size of optical systems. This is because the different monochromatic and chromatic aberrations in optical systems depend critically on the materials' properties [1]. In the previous chapter, I have shown that a key part of this potential is that nanocomposites allow for the design of highly efficient DOEs. However, the full potential of dispersion-engineered nanocomposites goes beyond this application. In fact, in Sect. 4.1, I have already demonstrated that the high refractive index of nanocomposites is highly useful for reducing spherical aberration. In this chapter, I now focus on using nanocomposites, or novel dispersion-engineered materials in general, for correcting chromatic aberrations, that is, the design of achromatic optical systems. My goal for this chapter is to develop advanced concepts for how such materials can be used. I emphasize that evaluating the full potential of dispersion-engineered materials is only possible through systematic optical design studies for different systems. As one promising application for novel dispersion-engineered materials, I investigate the potential of such materials for enabling a new generation of smartphone cameras in the final section of this chapter. The findings presented in this chapter will soon be published in a joint paper with my colleagues Markus Seesselberg, Hans-Jürgen Dobschal, and Vladan Blahnik.

6.1 Refractive Replacements for Diffractive Lenses

Diffractive lenses (DLs) are powerful tools for correcting chromatic aberrations in optical systems [2]. In Chap. 5, I have shown that DLs which can unlock this potential can be designed by tailoring the refractive indices such that the DOEs' phase profiles

D. Werdehausen, *Nanocomposites as Next-Generation Optical Materials*, Springer Series in Materials Science 316, https://doi.org/10.1007/978-3-030-75684-0_6

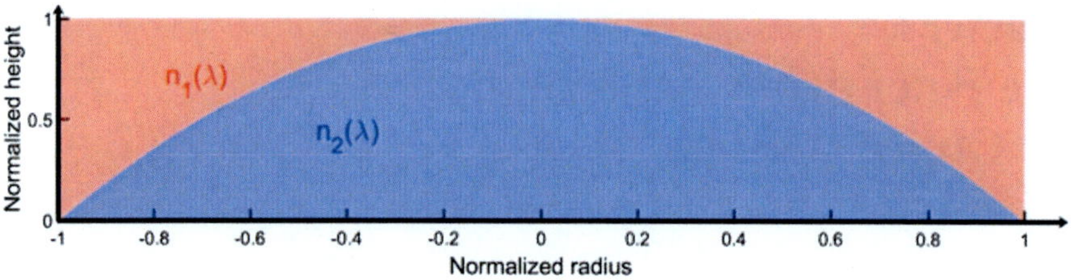

Fig. 6.1 Scheme of a two-layer refractive element that can replace a diffractive lens (DL) in a broadband optical system. To this end, the refractive indices $n_1(\lambda)$ and $n_2(\lambda)$ must be tailored according to my design rules (see Sect. 5.3). This makes the refractive element's phase profile essentially independent of the wavelength and it hence has the same effective Abbe number (ν_d^{eff}) and partial dispersion ($P_{g,F}^{eff}$) as a DL

become independent of the wavelength. In this section, I now demonstrate that this finding also opens up new possibilities for refractive components.

The effective Abbe number (ν_d^{dif}) and partial dispersion ($P_{g,F}^{dif}$) of DLs are derived based on the assumption that their phase profiles are independent of the wavelength (see Sect. 2.4.2). However, by applying my design rules from Sect. 5.3, it is also possible to design purely refractive elements whose phase profiles are independent of the wavelength. To demonstrate this possibility, I now consider the two-layer structure depicted in Fig. 6.1. Within the thin element approximation (TEA), this device's phase profile can be expressed as a function of the radial coordinate (r) as follows:

$$\phi(r) = \frac{2\pi}{\lambda} cr^2 (n_1(\lambda) - n_2(\lambda)) = \frac{2\pi}{\lambda} cr^2 \Delta n(\lambda), \tag{6.1}$$

where c quantifies the interface's curvature and I normalized the phase such that $\phi(r = 0) = 0$. As discussed in Sect. 5.3, my design rules for broadband DOEs yield refractive index differences that scale almost linearly with the wavelength ($\Delta n(\lambda) \propto \lambda$). For material combinations that fulfill this condition, the phase profile in (6.1) hence also becomes essentially independent of the wavelength. In this case, the two-layer refractive structure depicted in Fig. 6.1 consequently has the same effective Abbe number (ν_d^{eff}) and partial dispersion ($P_{g,F}^{eff}$) as a DL. This is because, for DLs, these parameters are directly derived from the assumption of a wavelength-independent phase profile (see Sect. 2.4.2). To confirm that this holds true, Fig. 6.2 depicts the focal shifts of a DL with a focal length of $f = 400$ mm and an equivalent refractive element that is composed of the same material combination as EG 1 (Table 5.1) as functions of the wavelength. These data were obtained using the optical design software "OpticStudio". Figure 6.2 shows that the DL and the refractive device indeed exhibit almost the same behavior. The slight difference at short wavelengths can be attributed to the finite thickness of the refractive device, which is not considered within the TEA. This finding that a two-layer refractive element can have the same effective Abbe number and partial dispersion as a DL shows that

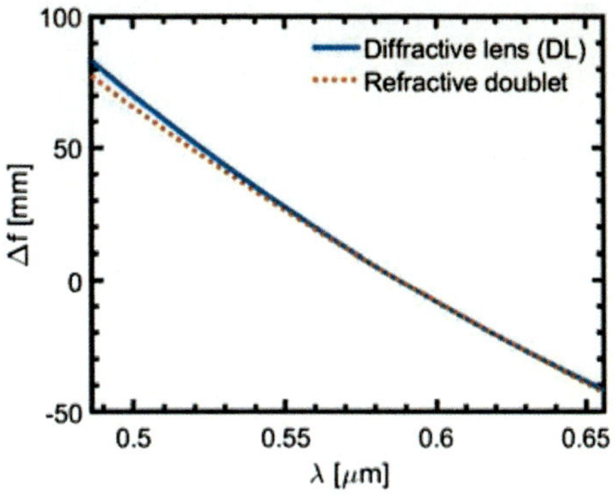

Fig. 6.2 Focal shift ($\Delta f(\lambda) = f(\lambda) - f(\lambda_d)$) as a function of the wavelength (λ) for a DL with a focal length of $f = 400$ mm and an equivalent two-layer refractive device. The small difference at short wavelengths can be attributed to the finite thickness of the refractive device

purely refractive elements made of dispersion-engineered materials can fully replace DLs in broadband optical systems. This approach has the immense advantage that refractive elements do not suffer from spurious diffraction orders. In fact, since these refractive elements have the same ν_d^{eff} and $P_{g,F}^{eff}$ as a DL, the results from Sect. 5.2.4 for the individual NAs of DLs (NA^{dif}) also apply to these devices. Therefore, these refractive DL replacements only have to provide little refractive power to correct the chromatic aberrations of broadband systems. As a result, only thin layers of dispersion-engineered materials are required. This also justifies the use of the TEA. However, I note that these layers still have to be significantly thicker than a DL. This implies that materials' absorption coefficients must be lower than for DLs to maintain a high transmittance.

6.2 Dispersion-Engineered Materials for Correcting Chromatic Aberrations

Since refractive replacements for DLs have the same effective Abbe number and partial dispersion as DLs, these refractive elements are also powerful tools for correcting chromatic aberrations in broadband optical systems. However, DLs are characterized by fixed values of ν_d^{dif} and $P_{g,F}^{dif}$, whereas dispersion-engineered nanocomposites can span a wide range of both these quantities (see Fig. 4.1). Therefore, the more powerful and general approach is to directly use the materials as degree of freedoms in the design process. This allows for optimizing the materials such that the performance of the system at hand is maximized rather than just replacing DLs.

To investigate how dispersion-engineered materials can be used for correcting chromatic aberration, I designed an apochromat (see Sect. 2.1.1) with a focal length of $f(\lambda_d) = 100$ mm and an aperture of $D = 12.5$ mm. As the first two materials for this system, I chose the standard optical glasses K7 and F2, which also define the normal line in the partial dispersion diagram (see Sect. 2.4.2 and Fig. 2.6). As the third material, I assumed a tunable optical material with variable values of ν_d as

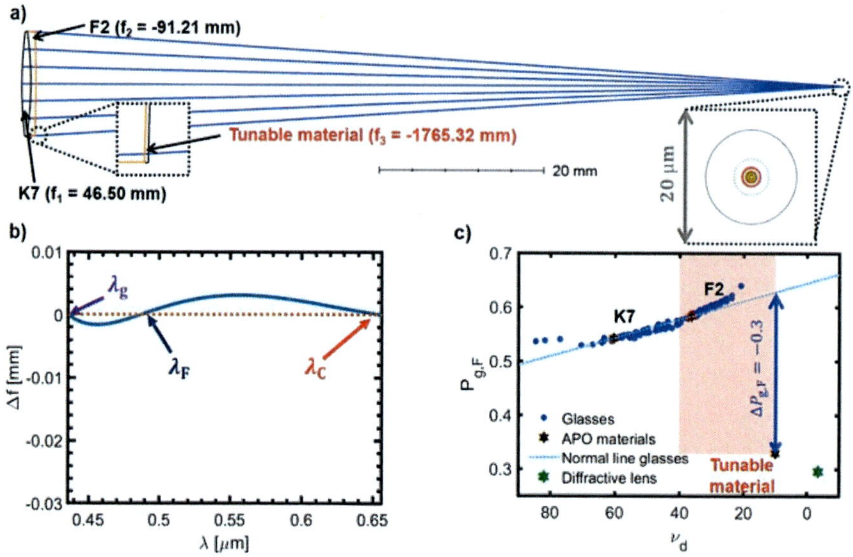

Fig. 6.3 **a** Layout of an apochromat composed of the optical glasses F2 and K7 as well as a thin layer of a tunable optical material. The apochromat has a focal length of $f(\lambda_d) = 100$ mm and an aperture of $D = 12.5$ mm. The black circle in the right inset denotes the Airy disk's diameter of $d_{Airy} = 5.7$ µm. The root mean square spot size obtained from ray-tracing is $d_{RMS} = 1.0$ µm. **b** Focal shift $\Delta f(\lambda) = f(\lambda) - f(\lambda_C)$ as a function of the wavelength to show that the apochromatization condition $f(\lambda_g) = f(\lambda_F) = f(\lambda_C)$ is fulfilled. **c** Partial dispersion diagram visualizing the ν_d and $P_{g,F}$ values of the apochromat's materials. The red area highlights the region within which the tunable material was optimized ($n_d = 1.6$). The dotted blue line denotes the normal line for optical glasses

well as $P_{g,F}$ but a fixed value of $n_d = 1.6$. Subsequently, I optimized the system's geometry and third material for a minimum spot size at all wavelengths (λ_g, λ_F, λ_d, and λ_C). In doing so, I constrained the tunable material to be located within the region in the partial dispersion diagram that is covered by nanocomposites (see Fig. 4.1b). Furthermore, I required the system to fulfill the apochromatization condition $f(\lambda_g) = f(\lambda_F) = f(\lambda_C)$. Figure 6.3a depicts the optimized system's layout. In addition, Fig. 6.3b presents its focal shift (Δf) as a function of the wavelength to show that the system is indeed apochromatic. Finally, Fig. 6.3c visualizes the materials' locations in the partial dispersion diagram. These figures demonstrate that apochromatization can be achieved by using just a thin layer of a dispersion-engineered material (see Fig. 6.3a, b). In fact, this layer only needs to be 18 µm thick. Accordingly, the values for the three elements' individual focal lengths, which are listed in Fig. 6.3a, highlight that the focal length of the thin layer is more than a factor of 17 larger than the overall focal length of the apochromat. Moreover, Fig. 6.3b shows that the maximum remaining color error ($\Delta f(\lambda)$) is below 5 µm, and the right inset in Fig. 6.3a visualizes that the system achieves diffraction limited performance at all wavelengths. In this inset, the diffraction limit is visualized by the solid black circle, which denotes the diameter of the Airy disk. Finally, the location of the tunable material in the partial dispersion diagram (Fig. 6.3c) shows that the tunable material converged to the bottom right corner of the region that was used for the optimization. This region is visualized by the red area in Fig. 6.3c and approximately corresponds to the range that is covered by nanocomposites in the partial dispersion diagram (Fig. 4.1b). The finding that the tunable material is located in this corner confirms that this material is characterized by anomalous values of ν_d and $P_{g,F}$ (Fig. 6.3c). In addition, this also explains why this layer can be very thin. This is because, as evident from (5.8), the amount of refractive power required from the third element in an apochromat is inversely proportional to the distance of its $P_{g,F}$ value to the line that connects the apochromats's other two materials in the partial dispersion diagram ($\Delta P_{g,F}$ in Fig. 6.3c). Furthermore, (5.8) also demonstrates that, if the first and second lenses' materials remain fixed, the third element's refractive power can be further reduced by minimizing its material's Abbe number. This indicates that it would also be possible to achieve a similar effect if $\Delta P_{g,F}$ is positive and sufficiently large, which corresponds to a location far above the normal line in Fig. 6.3c. In the appendix (Sect. A.9), I provide an additional design example to highlight that this is indeed true. These design examples for apochromats hence demonstrate that thin layers of novel dispersion-engineered materials are suitable for correcting chromatic aberration to a very high degree. To this end, the materials must be distinguished by anomalous partial dispersions and low Abbe numbers. This highlights that my approach of using the materials as continuous degrees of freedom within the entire range of properties that is covered by nanocomposites (see Fig. 4.1) opens up new possibilities for optical design.

To show that without tunable optical materials, the apochromat's performance is drastically impaired, I also designed a system using just optical glasses. For this benchmark system, I again used the optical glasses K7 and F2 as the first two materials but also used an optical glass as the third material. To obtain the best possible design in combination with K7 and F2, I allowed this final material to vary freely within the full Schott glass catalog [3]. The optimized system's layout is depicted in Fig. 6.4a. First, this figure shows that this system's lenses are significantly bulkier than in the case of the system that includes the dispersion-engineered material (Fig. 6.3a). Second, the spot diagram in the inset of Fig. 6.4a visualizes that the spot size is significantly increased and Fig. 6.4b demonstrates that the residual remaining color error is also drastically increased. Finally, Fig. 6.4c visualizes that the three optical glasses span a much smaller region in the partial dispersion diagram compared to the materials of the initial design that includes the dispersion-engineered material (Fig. 6.3c). Since the refractive power of an apochromat's elements can be minimized by maximizing the area of the triangle that is defined by the apochromats' three materials in the partial dispersion diagram [4], this explains why the individual lenses in this design must be much bulkier than before (Fig. 6.3). Traditionally, this area and hence the apochromat's performance could be increased by using so-called special glasses in the apochromat's first two lenses. The locations of such special glasses in the partial dispersion diagram deviate from the normal line (dotted blue line in Fig. 6.4c) [4]. However, such glasses are expensive and require intricate machining.

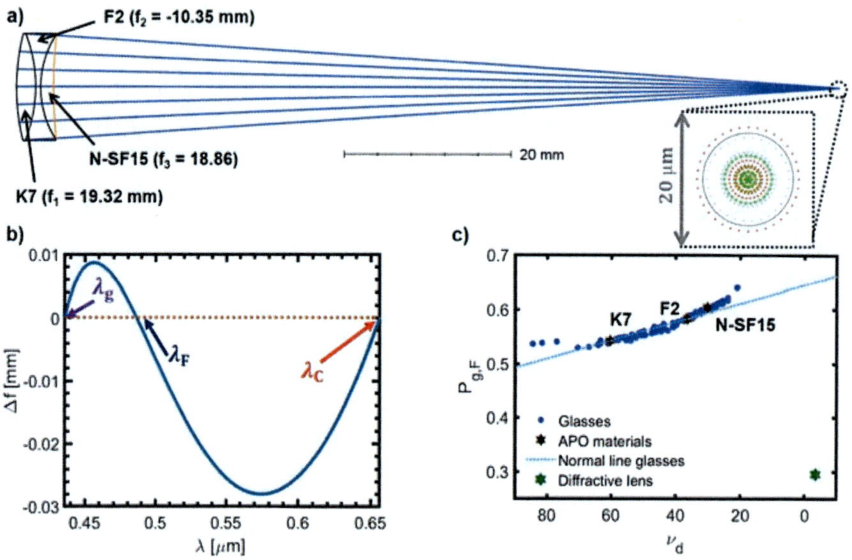

Fig. 6.4 **a** Layout of an apochromat composed of the optical glasses F2, K7, and N-SF15. The apochromat has a focal length of $f(\lambda_d) = 100$ mm and an aperture of $D = 12.5$ mm. The black circle in the right inset denotes the Airy disk's diameter of $d_{Airy} = 5.7$ μm. The root mean square spot size obtained from ray-tracing is $d_{RMS} = 3.3$ μm. **b** Focal shift $\Delta f(\lambda) = f(\lambda) - f(\lambda_C)$ as a function of the wavelength to show that the apochromatization condition $f(\lambda_g) = f(\lambda_F) = f(\lambda_C)$ is fulfilled. **c** Partial dispersion diagram visualizing the ν_d and $P_{g,F}$ values of the apochromat's materials

This is one of the key cost drivers of high-end optical systems. Furthermore, the locations of all glasses in the partial dispersion diagram in Fig. 6.4c demonstrates that the possibilities for maximizing the triangle's area with conventional optical glasses are limited. Therefore, my design examples for apochromats suggest that dispersion-engineered materials with anomalous values of ν_d and $P_{g,F}$ could replace special glasses. In fact, these examples also indicate that my approach of tailoring the materials to the system at hand could, at the same time, even increase the performance or decrease the size of optical systems. However, so far, materials with anomalous values of ν_d and $P_{g,F}$ have not been investigated in the scientific literature. My design examples highlight that, if such materials became available at reasonable prices, they would almost certainly lead to a paradigm shift in optical design.

6.3 Pushing the Limits of Smartphone Cameras

To conclude this chapter, I now investigate whether the concepts developed in the preceding chapters can have a significant impact on real-world consumer systems. For this analysis, I choose the example of smartphone cameras, because today they are probably the most widely used optical systems in the world. In fact, these cameras have reached an astonishing quality that was unimaginable less than a decade ago. This is mainly due to the heavy use of highly complicated aspheric lenses [5]. However, because of the limited thickness of modern smartphones, increasing the focal lengths of such cameras modules is a highly challenging task. This is because realizing focal lengths that are longer than the system's total length (TL) requires an increasing amount of refractive power within the system as the focal length is increased [1, 6]. This, in turn, increases the aberrations. Therefore, the currently immense demand for telephoto lenses in smartphones is difficult to satisfy. For example, the so-called telephoto-module of the Iphone Xs has a focal length of 6 mm [7]. Taking the sensor's size into consideration, this corresponds to a focal length of merely 52 mm on a full-frame camera [7]. This focal length is below the threshold of 75 mm beyond which a lens is usually regarded as a telephoto system [1]. Furthermore, because of their compact size, smartphone camera modules also have much smaller sensors and pixel sizes than conventional cameras. Today pixel pitches below 1 μm are typically used [8]. Therefore, a sufficiently large aperture is essential to reduce the diffraction limit to a level at which a single pixel can be resolved. For example, the diffraction limit ($d_{\mathrm{lim}} = \frac{\lambda}{2\mathrm{NA}}$ [1]) at a typical f-number of $N_{\mathrm{f}} = 2.8$ ($\mathrm{NA} \approx \frac{1}{2N_{\mathrm{f}}} = 0.18$) and a wavelength of $\lambda_{\mathrm{d}} = 0.588\ \mu\mathrm{m}$ is $d_{\mathrm{lim}} = 1.65\ \mu\mathrm{m}$. A single pixel on a sensor with a pixel pitch below 1 μm is consequently already significantly smaller than the diameter of the airy disk. In addition, decreasing the f-number increases the amount of light that reaches the sensor and hence improves the system's performance in low light conditions [6]. Approaches that allow for simultaneously increasing the focal lengths and decreasing the f-number of smartphone cameras can hence be expected to have an immense impact. For this reason, I now investigate whether novel dispersion-engineered material can push smartphone cameras beyond the current limits. To this end, I here provide an overview over an optical design study based on which we recently filed a patent application [9] and that will soon be published. Note that, for a fixed sensor size (h), the optical system's focal length directly determines the system's field of view (FOV) according to $\mathrm{FOV} = 2 \arctan\left(\frac{h}{2f}\right)$ [6]. Increasing a system's focal length hence goes hand in hand with decreasing its FOV.

As the benchmark for our designs, we chose a state-of-the-art design from an Apple patent [10]. Figure 6.5a demonstrates that this system consists of five refractive lenses with complicated aspheric shapes. As listed in Table 6.1, the corresponding materials were only specified in terms of their refractive index at the d-line (n_{d}) and Abbe number (ν_{d}). Note that the n_{d} values of these materials are significantly higher than those of readily commercially available polymers. Furthermore, the system was optimized for object distances between infinity and 500 mm. To visualize the system's performance, Fig. 6.5b depicts its modulation transfer function (MTF) at an object distance of infinity as a function of the field angle at three different spatial frequencies. For a sensor with a typical pixel pitch of $d_{\mathrm{pixel}} = 0.9\ \mu\mathrm{m}$, these frequencies correspond to $\frac{\mathrm{Nyq}}{2}$, $\frac{\mathrm{Nyq}}{4}$, and $\frac{\mathrm{Nyq}}{8}$, where Nyq is the sensor's Nyquist frequency ($\mathrm{Nyq} = \frac{1}{2d_{\mathrm{pixel}}}$). These spatial frequencies are commonly used to quantify the perfor-

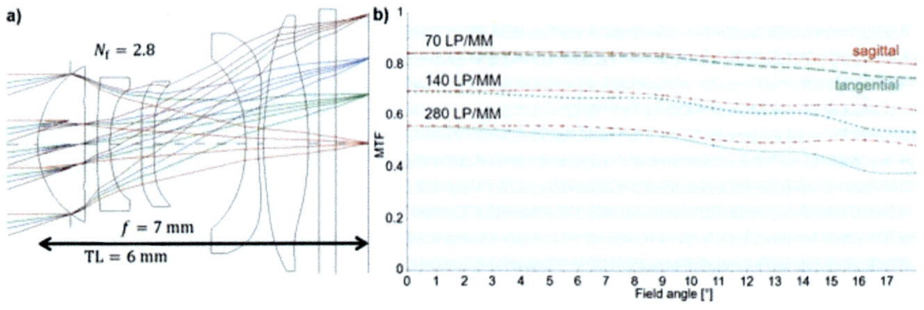

Fig. 6.5 **a** Layout of a state-of-the-art smartphone telephoto system from an Apple patent [10]. The lens has a focal length of $f = 7$ mm, a f-number of $N_f = 2.8$, a total length of TL = 6 mm, and a field of view of FOV = 36°. **b** MTF of the system at an object distance of infinity as a function for the field angle (the AOI) for three different spatial frequencies (LP/MM denotes the line pairs per millimeter). A detailed overview over the system's performance is provided in the appendix (Sect. A.10)

Table 6.1 Materials of which the lenses in the smartphone designs discussed in this section are composed. The benchmark design was extracted from an Apple patent [10] and is depicted in Fig. 6.5. The materials on which this Apple design is based were only specified in terms of their n_d and ν_d values. In contrast, the double DL design is based on readily available commercial polymers (Zeonex: $n_d = 1.53$ and $\nu_d = 56$; PC: $n_d = 1.58$ and $\nu_d = 27.8$). This design is depicted in Fig. 6.6

Design	1. Lens	2. Lens	3. Lens	4. Lens	5. Lens
Benchmark	$n_d = 1.54$, $\nu_d = 56.1$	$n_d = 1.63$, $\nu_d = 23.3$	$n_d = 1.63$ $\nu_d = 23.3$	$n_d = 1.54$, $\nu_d = 56.1$	$n_d = 1.54$, $\nu_d = 56.1$
Double DL	Zeonex	PC	PC	Zeonex	PC

mance of an imaging chain consisting of both a lens and a sensor [5]. I provide a full overview over the system's performance in the appendix (Sect. A.10).

To investigate whether novel materials can push smartphone cameras beyond their current limits, we designed a system that includes DLs. On the one hand, this allows me to investigate whether the unprecedented high efficiencies of nanocomposite-enabled DLs make novel design for smartphone cameras possible. On the other hand, as shown in the previous section, DLs can also be replaced by thin layers of dispersion-engineered materials. However, including additional layers and optimizing materials complicates the design process. Instead, DLs can readily be incorporated on existing surfaces. In fact, because DLs are intrinsically characterized by anomalous values of ν_d^{dif} and $P_{g,F}^{dif}$, they are ideally suited for evaluating both the benefits of DLs themselves and the general potential of materials with anomalous values of ν_d and $P_{g,F}$. Note that the DLs were treated within the "optical design perspective" (see Sect. 5.2.1), which assumes ideal DLs with efficiencies of 100%. Therefore, I below again use the DLs' minimal segment widths (Λ_{min}) to evaluate whether nanocomposite-enabled DLs allow for keeping the amount of stray light in spurious diffraction orders in the system at a negligible level (see Sect. 5.2.4). First, by

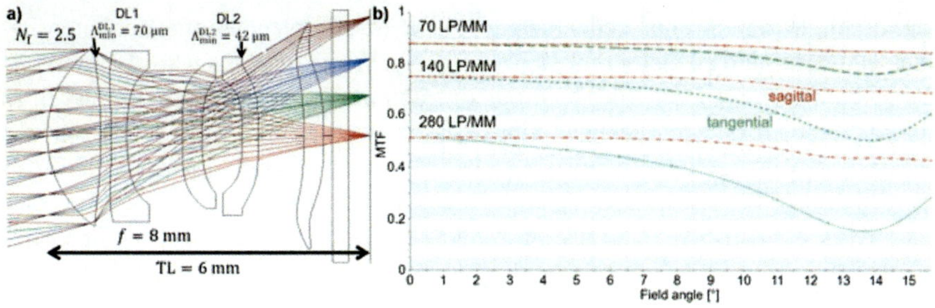

Fig. 6.6 **a** Layout of smartphone telephoto system with two diffractive lenses (DLs). The system has a focal length of $f = 8$ mm, a f-number of $N_f = 2.5$, a total length of TL $= 6$ mm and a field of view of FOV $= 31.4^o$. **b** MTF of the system at an object distance of infinity as a function for the field angle (the AOI) for three different spatial frequencies. A detailed overview over the system's performance is provided in the appendix (Sect. A.10)

systematically designing different systems, we found that the highest benefit can be achieved by incorporating two DLs at different locations in the system. Accordingly, Fig. 6.6a depicts the layout of a hybrid system with two DLs. As highlighted in the figure, this system has both an increased focal length of $f = 8$ mm and a decreased f-number of $N_f = 2.5$. This focal length is equivalent to $f = 77$ mm on a full frame camera and is hence within the range of typical telephoto lenses. Compared to the benchmark design, the system's focal length was consequently increased by 11%, and, at the same time, its f-number was decreased by 14%. Since the total length of TL $= 6$ mm remained unchanged, this corresponds to a significant improvement. In fact, the decrease of the f-number by 14% corresponds to an increase of the system's etendue, that is, the amount of light that is delivered to the sensor by more than 26%. Therefore, the system would allow for a significantly increased performance in low light conditions. Moreover, the comparison of Figs. 6.5b and 6.6b shows that the MTFs of the two systems are similar in terms of the overall magnitude at the different spatial frequencies. This demonstrates that the focal length was increased and the f-number decreased without reducing the system's performance at the object distance of infinity. In addition, as opposed to the Apple design, this was achieved using readily available polymers that have lower refractive indices than the materials assumed in the Apple patent [10]. I again present a detailed overview over the system's performance in the appendix (Sect. A.10). In addition, the two DLs' minimal segment widths (Λ_{min}), which are specified in Fig. 6.6a, confirm that the DLs must only provide little optical power (see Sect. 5.2.4). Specifically, the values of $\Lambda_{min}^{DL1} = 70\ \mu m$ and $\Lambda_{min}^{DL2} = 42\ \mu m$ are well within the range for which nanocomposite-enabled EGs can maintain a high performance (see Sect. 5.1). Moreover, the ray bundles for different field angles become spatially separated towards the end of the system (see Fig. 6.6a). Therefore, the second DL, which has the smaller value of Λ_{min}, can be locally adjusted for a narrow range of angles of incidence (AOIs). The range of AOIs across which nanocomposite-enabled DLs maintain a high performance (see Sect. 5.2.4) hence covers the entire range of AOIs that is required within the system. These findings highlight that the unprecedented high diffraction efficiencies of nanocomposite-enabled DLs would allow for realizing such designs with two DLs while maintaining the amount of stray light from spurious diffraction orders at a negligible level. In contrast, until now, such designs with two DLs were unthinkable because of the generally low diffraction efficiencies and small region of grating periods and AOIs for which the state-of-the-art solutions maintain reasonably high efficiencies (see Sect. A.3). Specifically, as discussed in Sect. 5.2.4, the low diffraction efficiencies of the state-of-the-art solutions would drastically affect the system's MTF and lead to colorful flares as well as ghost images. Furthermore, as shown in the previous section, nanocomposites or other novel dispersion-engineered materials allow for completely replacing DLs by thin refractive layers. This would have the immense advantage that stray light in spurious diffraction orders would not be an issue at all. However, both the design of highly efficient EGs and their refractive replacements require novel materials with anomalous values of ν_d and $P_{g,F}$ (see Sects. 5.3 and 6.1). The design with two DLs in Fig. 6.6 consequently indicates that such materials indeed have the potential to become an enabling technology for a next generation of optical systems. In fact, I here focused on the potential of DLs. However, exploiting the full dispersion-engineering capabilities of nanocomposites, also for the lenses itself, could potentially push the limits even further.

6.3.1 Nanocomposites Could Change the Way How Chromatic Aberrations are corrected and Could Enhance Real-World Optical Systems

In this chapter, I have first shown that DOEs in broadband optical systems can be replaced by purely refractive doublets made of materials, which also allow for the design for highly efficient DOEs. According to my design rules for highly efficient broadband DOEs from Sect. 5.3, fulfilling this condition to a high degree requires at least one material that is distinguished by an anomalous partial dispersion, that is, a value of $P_{g,F}$ that significantly deviates from those of normal materials. Building on this finding, I have then shown that thin layers of such materials, with thicknesses in the order of 100 μm, can be used in apochromatic systems to correct chromatic aberrations to a very high degree. Therefore, if such materials become available at reasonable cost, this could completely change the way how chromatic aberrations in optical systems are corrected. Finally, I have demonstrated for the specific example of a smartphone telephoto camera that the concepts I developed in this book indeed hold a high potential for significantly improving real-world optical systems. For this example, I have focused solely on nanocomposite-enabled DOEs. In the future, additional design studies will be required to determine by how much the limits of such an other systems can be pushed even further by additionally using the dispersion-engineering capabilities also for the refractive lenses.

References

1. H. Gross, W. Singer, M. Totzeck, F. Blechinger, B. Achtner, *Handbook of Optical Systems*, vol. 1 (Wiley-VCH, Berlin, 2005)
2. T. Stone, N. George, Hybrid diffractive-refractive lenses and achromats. Appl. Opt. **27**(14), 2960–71 (1988)
3. Schott, Optical Glass 2020. Technical Report. Schott AG (2020)
4. 3R. I. Mercado. Design of apochromats and superachromats, in *Lens Design: A Critical Review*, ed by W.J. Smith, vol. 10263 (International Society for Optics and Photonics. SPIE, 1992), pp. 269–295
5. T. Steinich, V. Blahnik, Optical design of camera optics for mobile phones. English. Adv. Opt. Technol. **1**(1–2), 51–58 (2012)
6. R. Kingslake, *Lenses in Photography: The Practical Guide to Optics for Photographers* (Barnes, 1963)
7. F. Gallagher, What is the focal length of an iPhone Camera and Why Should I Care (2019). https://improvephotography.com/55460/what-is-the-focal-length-of-an-iphone-camera-and-whyshould-i-care/.Webpage. Accessed 22 April 2020
8. R.J. Gove, 7—CMOS image sensor technology advances for mobile devices, in *High Performance Silicon Imaging*, 2nd edn., ed. by D. Durini (Woodhead Publishing Series in Electronic and Optical Materials. Woodhead Publishing, 2020), pp. 185–240
9. D. Werdehausen, H.-J. Dobschal, V. Blahnik, Kompaktes Teleobjektiv mit diffraktivem optischen Element. G DE102020105201.4 (2020)
10. R.I. Mercado, Small form factor telephoto camera. Patent US2015/0116569 (2015)

Chapter 7
Summary and Outlook

In this book, I have investigated the fundamental properties of optical nanocomposites and their potential as next-generation optical materials. To summarize how my findings bridge different gaps between fundamental research and practice, I now address each of the four main questions that I set out to answer in the introduction.

Since nanocomposites are composed of distinct scatterers that are much larger than the atoms or molecules in conventional optical materials, my first key question was: **Can nanocomposites be used as bulk optical materials?** In Chap. 3, I showed that this is indeed possible using large-scale T-matrix simulations of nanoparticle distributions that are composed of hundreds of thousands of individual scatterers. However, this analysis also demonstrated that nanocomposites can only be used as bulk optical materials within the "homogeneous regime". This regime is reached if the nanoparticles are smaller than one hundredth of the wavelength. Furthermore, to reach this regime, it is essential that the nanoparticles are well-dispersed and well-blended into the host matrix. In general, the fact that my large-scale simulations were able to reach the homogeneous regime shows that my approach allows for modelling homogeneous optical bulk materials at the single scatterer level. Furthermore, I demonstrated that, if the nanoparticles' sizes in nanocomposites are increased beyond one hundredth of the wavelength, the materials will transition into the "restricted effective medium regime". In this regime, the concept of an effective refractive index breaks down. Specifically, (1) the effective refractive indices are no longer well-defined but rather exhibit strong fluctuations. (2) There is no effective refractive index that can fully quantify the materials' optical responses, that is, different values are required depending on the measure that is used (e.g. reflection or transmission). (3) Incoherent scattering plays a key role as an attenuation mechanism. This mechanism distorts a coherent beam of light by successively redirecting a portion of its power flux in other directions. In imaging applications, this would lead to stray light and hence an unacceptable loss of contrast.

D. Werdehausen, *Nanocomposites as Next-Generation Optical Materials*, Springer Series in Materials Science 316, https://doi.org/10.1007/978-3-030-75684-0_7

Building on the finding that it is possible to use nanocomposites as bulk optical materials, my second main research question subsequently was: **What is the potential of nanocomposites as optical materials?** In Chap. 4, I showed that nanocomposites in the homogeneous regime can serve as a platform of dispersion-engineered materials. To this end, I used a systematic analysis based on the Maxwell-Garnett-Mie effective medium theory (EMT) and experimental data for different materials. This approach demonstrated that combining different nanoparticle materials within a host matrix allows for controlling the magnitude and dispersion of the nanocomposites' effective refractive index within wide regions. This unlocks the materials as continuous degrees of freedom that can be optimized for the application at hand. In addition, I provided an initial experimental proof of principle for the concept of nanocomposite-enabled optical elements. To this end, I used prototypical micro-optical lenses that were 3D printed out of different nanocomposites. A systematic characterization of the imaging properties of these lenses has then shown that they indeed exhibit the behavior that is expected from the nanocomposites' properties. Finally, both the experimental data and my large-scale numerical simulations showed that the Maxwell-Garnett-Mie EMT is a highly accurate tool for the design of nanocomposites. Therefore, I subsequently used this EMT to develop general concepts for how nanocomposites can enable a next generation of optical systems.

Because diffractive optical elements (DOEs) have a high potential for broadband applications, which has so far largely remained unfulfilled, my third main research question was: **Do nanocomposites allow for the design of highly efficient diffractive optical elements for broadband applications?** In Chap. 5, I demonstrated that this is indeed possible. Specifically, I showed that the use of nanocomposites in both layers of échelette-type gratings (EGs) enables the design of DOEs that achieve an average diffraction efficiency in the visible spectrum (400 nm $\leq \lambda \leq$ 800 nm) of over 99.9%. To this end, the nanocomposites' dispersion must be engineered such that the EGs' phase profiles become close to independent of the wavelength. Subsequently, I showed using an all-system analysis that diffractive lenses (DLs) based on my concept of nanocomposite-enabled EGs allow for unlocking the full potential of DLs even for high-numerical-aperture (high-NA) broadband imaging systems. In addition, this analysis yielded that the individual NA of a DL in a broadband imaging system is generally between one and two orders of magnitude smaller than the overall system's NA. This shows that high-NA "flat optics" are not required for broadband imaging systems. Finally, I generalized my concept of using dispersion engineering to achieve high diffraction efficiencies. To this end, I developed a general formalism and rules for how the dispersion must be tailored. This design formalism is independent of the material platform. Furthermore, I demonstrated that my design rules are valid for other DOEs that rely on propagation delays and hence also apply to metalenses.

To reach a final conclusion, my fourth research question was: **Can nanocomposites provide significant benefits for optical systems that outweigh their increased complexity?** To address this question, I developed general concepts for how novel dispersion-engineered materials can enable a next generation of optical systems in Chap. 6. First, I showed that dispersion-engineered materials, such as nanocomposites, allow for replacing DLs in broadband optical systems by purely refractive two-layer elements. This can generally be achieved by tailoring the two layers' refractive indices according to my design rules for broadband DOEs. Since these rules render the elements' phase profiles almost independent of the wavelength, such two-layer elements have the same effective Abbe number (ν_d) and partial dispersion ($P_{g,F}$) as DLs. This approach of replacing DLs by two-layer refractive elements has the immense advantage that it allows for exploiting the benefits of DLs without suffering from the drawback of stray light in spurious diffraction orders. Subsequently, I generalized this concept and showed that thin layers with thicknesses around 100 μm can be used to correct chromatic aberrations in broadband systems to a high degree. For this purpose, materials that have low Abbe numbers and highly anomalous partial dispersions are required. In addition, this approach allows for optimizing the materials such that the aberrations of a specific system are minimized. If such materials become available at reasonable cost, this could therefore completely change the way how chromatic aberrations in broadband optical systems are corrected. Finally, I investigated the potential of novel dispersion engineered materials for real-world optical systems on the example of a smartphone telephoto camera. I chose this example because there currently is an immense demand for smartphone cameras with longer focal lengths, which is difficult to fulfill because of the limited thicknesses of modern smartphones. To evaluate whether the approaches developed in the preceding chapters allow for pushing such systems beyond their current limits, I presented a design for a telephoto smartphone camera which includes two DLs. This design showed that incorporating two nanocomposite-enabled DLs in a smartphone telephoto camera allowed for increasing the focal length by 11% and decreasing the f-number by 14% compared to a benchmark system from an Apple patent [1]. In fact, utilizing the full dispersion-engineering capabilities of nanocomposites and not only nanocomposite-enabled DLs could possibly push these limits even further. This example illustrated that novel dispersion-engineered materials could enable a next generation of optical systems.

Looking ahead

In the future, unlocking the potential of the concepts developed in the preceding chapters for real-world applications will require immense interdisciplinary efforts. In doing so several major challenges must be overcome. Furthermore, my work opens up new research topics for fundamental research. To conclude this book, I now provide my perspective on these challenges and directions for future work.

First, my approach of numerically modeling optical materials at the single scatterer level can be readily generalized. In my work, I focused on random nanocomposites that are composed of spherical nanoparticles at volume fractions of up to $f = 35\%$, because this range is readily experimentally accessible [2]. However, other types of scatterers including atoms, molecules [38], and nanoparticles with other shapes [3–8] can be readily modeled within the same framework. Furthermore, other kinds of statistical scatterer distributions can also be investigated. In fact, both random distributions themselves [9–13] and the transition region between disordered and ordered distributions [14–17] possess a fascinating complexity. Therefore, using my approach to investigate other kinds of scatterers and different statistical scatterer distributions will allow for gaining a much deeper understanding of optical materials. As I demonstrated in Chap. 3, such investigations are also essential to guide the design of materials with novel properties.

Second, the finding that the nanoparticles in bulk optical nanocomposites must be smaller than one hundredth of the wavelength opens up new routes for tailoring the effective refractive indices of nanocomposites that have so far not been explored. This is because nanoparticles with sizes in the low single-digit nanometer range are located in the quantum dot regime. In this regime, the nanoparticles' electronic transitions can be tailored by varying their surface and size [18, 19]. Since any material's refractive index is determined by the absorption lines, that is, the optical resonances within the system [20], incorporating quantum dots into nanocomposites provides a powerful new degree of freedom for engineering the effective refractive indices. This expands the dispersion-engineering capabilities of nanocomposites even beyond the results presented in Chap. 4, since I here assumed the nanoparticles' permittivities to be equal to their materials' bulk values. Specifically, starting from the bulk values, quantum dots enable tailoring the nanoparticles' permittivities within a certain region. In my opinion, this is a promising approach that warrants future research.

Third, quantifying the benefits of nanocomposites, or novel dispersion-engineered materials in general, for optical systems requires systematic optical design studies. Such studies are also essential to identify the systems that benefit the most from such materials. In fact, my general findings suggest that two types of optical systems will benefit the most: (1) Systems that rely on materials that cover only limited range of refractive indices at the d-line (n_d), Abbe numbers (ν_d), and partial dispersions ($P_{g,F}$). The most prominent examples for such systems certainly are consumer systems like smartphone cameras that rely solely on polymers. (2) Systems that suffer heavily from chromatic aberrations. Mechanistically, it can be expected that such systems could benefit particularly from the use of DLs or materials with anomalous partial dispersions. This suggests that systematic optical design studies should initially focus on these two types of systems.

Fourth, materials with anomalous partial dispersions have so far not been in the focus of research efforts. As I demonstrated in Chap. 6, the key goal for the design of such materials should be to maximize the difference of their partial dispersions from those of conventional materials ($\Delta P_{g,F}$) and minimize their Abbe numbers. In fact, the range of Abbe numbers and partial dispersions that is accessible with conventional optical materials is mostly determined by the requirement that these materials must have negligible absorption [20]. However, I have shown that materials with anomalous partial dispersions must only maintain a high transmittance for thicknesses around 100 μm. Therefore, there is much more freedom in the design of such materials compared to conventional optical materials. This follows from the fact that dispersion is intrinsically connected to absorption (see Sect. 2.1.1). This also illustrates that the key to achieving anomalous partial dispersions is introducing additional resonances close to the visible spectrum into the material. A fascinating fundamental research question consequently is how much dispersion can be introduced into a material while still maintaining a high transmittance at thicknesses around 100 μm. Furthermore, a key question is whether such materials should be realized as nanocomposites or if dopants on the molecular level should be used to introduce additional resonances. Both these approaches have, for example, been pursued for the design of high-refractive-index polymers [21]. In this context, tailoring the absorption lines of quantum dots in nanocomposites might provide unique possibilities in the future. In addition, a fundamental advantage of nanocomposites is that crystalline nanoparticle materials generally have much higher atomic densities than amorphous materials [20]. Since increasing the atomic densities allows for increasing the refractive index [20], this suggests that nanocomposites are the most promising approach when it comes to maximizing the effective refractive index. Especially if the goal also is to achieve a high Abbe number, that is, a low amount of dispersion.

Finally, most of the materials investigated in this book have already been synthesized in research works [2, 21–30, 30–37]. However, transferring these approaches from research into mass-produced consumer applications is a major challenge. Specifically, I have shown that the key task for nanocomposites is to (1) synthesize nanoparticles with sizes below 4 nm; (2) design surface functionalizations for the nanoparticles that enable a perfect blending of the nanoparticles into different host matrices; (3) the materials must be designed such that they can be integrated into standard production processes, and (4) be relatively cheap. Integrating novel materials into consumer products will certainly also involve adjusting current production chains. Therefore, my conclusion is that, if novel dispersion-engineered materials become available at reasonable prices, they will most likely lead to a paradigm shift in optical design. However, there are still major challenges that need to be overcome to unlock this potential. These challenges can only be overcome through immense interdisciplinary efforts. However, regarding these challenges, I draw hope from the fact that, because of immense technological advances, optical glasses today have reached an astonishing quality that was also unimaginable a few decades ago [20].

References

1. R.I. Mercado, Small form factor telephoto camera. Patent US2015/0116569 (2015)
2. P. Tao, Y. Li, A. Rungta, A. Viswanath, J. Gao, B.C. Benicewicz, R.W. Siegel, L.S. Schadler, TiO2 nanocomposites with high refractive index and transparency. J. Mater. Chem. **21**(46), 18623–18629 (2011)
3. X.G. Santiago, M. Hammerschmidt, S. Burger, C. Rockstuhl, I. Fernandez-Corbaton, L. Zschiedrich, Decomposition of scattered electromagnetic fields into vector spherical wave functions on surfaces with general shapes. Phys. Rev. B **99**(4), 045406 (2019)
4. D. Theobald, A. Egel, G. Gomard, U. Lemmer, Plane-wave coupling formalism for T-matrix simulations of light scattering by nonspherical particles. Phys. Rev. B **96**(3), 033822 (2017)
5. M.I. Mishchenko, L. Liu, D.W. Mackowski, B. Cairns, G. Videen, Multiple scattering by random particulate media: exact 3D results. Opt. Exp. **15**(6), 2822–2836 (2007)
6. M.I. Mishchenko, L.D. Travis, A. Macke, Scattering of light by polydisperse, randomly oriented, finite circular cylinders. Appl. Opt. **35**(24), 4927–4940 (1996)
7. E.A. Muljarov, T. Weiss, Resonant-state expansion for open optical systems: generalization to magnetic, chiral, and bi-anisotropic materials. Opt. Lett. **43**(9), 1978 (2018)

8. T. Weiss, E.A. Muljarov, How to calculate the pole expansion of the optical scattering matrix from the resonant states. Phys. Rev. B **98**(8) (2018)
9. R.M. Ziff, S. Torquato, Percolation of disordered jammed sphere packings. J. Phys. A: Math. Theor. **50**(8), 085001 (2017)
10. S. Torquato, Perspective: Basic understanding of condensed phases of matter via packing models. J. Chem. Phys. **149**(2), 020901 (2018)
11. S. Atkinson, F.H. Stillinger, S. Torquato, Existence of isostatic, maximally random jammed monodisperse hard-disk packings. Proc. Natl. Acad. Sci. U. S. A. **111**(52), 18436 (2014)
12. L. Pattelli, A. Egel, U. Lemmer, D.S. Wiersma, Role of packing density and spatial correlations in strongly scattering 3D systems. Optica **5**(9), 1037–1045 (2018)
13. M.A. Klatt, S. Torquato, Characterization of maximally random jammed sphere packings. III. Transport and electromagnetic properties via correlation functions. Phys. Rev. E **97**(1), 012118 (2018)
14. S. Torquato, F.H. Stillinger, Controlling the short-range order and packing densities of many-particle systems. J. Phys. Chem. B **106**(33), 8354–8359 (2002)
15. M.V. Rybin, D.S. Filonov, K.B. Samusev, P.A. Belov, Y.S. Kivshar, M.F. Limonov, Phase diagram for the transition from photonic crystals to dielectric metamaterials. Nat. Commu. **6**, 10102 (2015)
16. M. Rechtsman, A. Szameit, F. Dreisow, M. Heinrich, R. Keil, S. Nolte, M. Segev, Amorphous photonic lattices: band gaps, effective mass, and suppressed transport. Phys. Rev. Lett. **106**(19), 193904 (2011)
17. L. Shi, Y. Zhang, B. Dong, T. Zhan, X. Liu, J. Zi, Amorphous photonic crystals with only short-range order. Adv. Mater. **25**(37), 5314–5320 (2013)
18. S. Chandra, S.H. Pathan, S. Mitra, B.H. Modha, A. Goswami, P. Pramanik, Tuning of photoluminescence on different surface functionalized carbon quantum dots. RSC Adv. **2**(9), 3602–3606 (2012)
19. J.M. Garcia, T. Mankad, P.O. Holtz, P.J. Wellman, P.M. Petroff, Electronic states tuning of InAs self-assembled quantum dots. Appl. Phys. Lett. **72**(24), 3172–3174 (2020)
20. P. Hartmann, R. Jedamzik, S. Reichel, B. Schreder, Optical glass and glass ceramic historical aspects and recent developments: a Schott view. Appl. Opt. **49**(16), D157-D176 (2010)
21. J.-G. Liu, M. Ueda, High refractive index polymers: fundamental research and practical applications. J. Mater. Chem. **19**(47), 8907–8919 (2009)
22. M. Feuillade, G. Cantagrel, Liquid polymerizable composition comprising mineral nanoparticles and its use to manufacture an optical article. Patent US20150203710 (2015)
23. S. Li, M. Meng Lin, M.S. Toprak, D.K. Kim, M. Muhammed, Nanocomposites of polymer and inorganic nanoparticles for optical and magnetic applications. Nano Rev. **1**(1), 5214 (2010)
24. H.K. Schmidt, Sol-gel nanocomposites as functional optical materials, in *Sol-Gel Optics II*, ed. by J.D. Mackenzie, vol. 1758 (International Society for Optics and Photonics. SPIE, 1992), pp. 396–402
25. N. Nakashima, High refractive index glass compositions. Patent US4082427 (1978)
26. S. Monickam, D. Peters, G. Cooper, Z. Chen, Nanocomposite formulations for optical applications. Patent US20180223107 (2018)
27. J.L.H. Chau, Y.-M. Lin, A.-K. Li, W.-F. Su, K.-S. Chang, S.L.-C. Hsu, T.-L. Li, Transparent high refractive index nanocomposite thin films. Mater. Lett. **61**(14–15), 2908–2910 (2007)
28. P.T. Chung, C.T. Yang, S.H. Wang, C.W. Chen, A.S. Chiang, C.-Y. Liu, ZrO2/epoxy nanocomposite for LED encapsulation. Mater. Chem. Phys. **136**(2–3), 868–876 (2012)
29. S. Lee, H.-J. Shin, S.-M. Yoon, D.K. Yi, J.-Y. Choi, U. Paik, Refractive index engineering of transparent ZrO2-polydimethylsiloxane nanocomposites. J. Mater. Chem. **18**(15), 1751–1755 (2008)
30. H. Liu, X. Zeng, X. Kong, S. Bian, J. Chen, A simple two-step method to fabricate highly transparent ITO/polymer nanocomposite films. Appl. Surface Sci. **258**(22), 8564–8569 (2012)
31. C. Lü, Z. Cui, Z. Li, B. Yang, J. Shen, High refractive index thin films of ZnS/polythiourethane nanocomposites. J. Mater. Chem. **13**(3), 526–530 (2003)

32. C. Lü, Z. Cui, Y. Wang, Z. Li, C. Guan, B. Yang, J. Shen, Preparation and characterization of ZnS-polymer nanocomposite films with high refractive index. J. Mater. Chem. **13**(9), 2189–2195 (2003)
33. C. Lü, C. Guan, Y. Liu, Y. Cheng, B. Yang, PbS/polymer nanocomposite optical materials with high refractive index. Chem. Mater. **17**(9), 2448–2454 (2005)
34. C. Lü, B. Yang, High refractive index organic-inorganic nanocomposites: design, synthesis and application. J. Mater. Chem. **19**(19), 2884–2901 (2009)
35. G. Cooper, W. Xu, Z. Chen, High refractive index nanocomposite layer. Patent US10144842. (2018)
36. A. Garito, Y.-L. Hsiao, R. Gao, J. Zhu, B. Thomas, A. Panackal, J. Sharma, R. Gao, Optical polymer nanocomposites. Patent US20030175004 (2003)
37. T. Ogata, R. Yagi, N. Nakamura, Y. Kuwahara, S. Kurihara, Modulation of polymer refractive indices with diamond nanoparticles for metal-free multilayer film mirrors. ACS Appl. Mater. Interfaces **4**(8), 3769–72 (2012)
38. I. Fernandez-Corbaton, D. Beutel, C. Rockstuhl, A. Pausch, W. Klopper, Computation of electromagnetic properties of molecularensembles. Chem. Phys. Chem **21**, 878 (2020)

Appendix

A.1 Numerical Modeling—Additional Data

In this section, I present supplementary data for my analysis of the transition between homogeneous and heterogeneous optical materials in Chap. 3. This section was adapted from [1].

A.1.1 The Role of the Placement Procedure

To demonstrate the strong influence of the particle placement procedure, I here use two alternative procedures. This also allows me to gain additional insights into how changes in the statistical properties of the nanoparticle distributions affect the effective refractive indices.

The two alternative placement procedures are schematically shown in Fig. A.1. For the first additional placement procedure (procedure II), I force the nanoparticles to be fully located within the cuboid that encompasses the nanoparticles (Fig. A.1a). For the second additional procedure (procedure III), I instead only require the nanoparticles' centers to be inside the cuboid (Fig. A.1b). However, in contrast to the placement procedure used throughout Chap. 3 (procedure I), I do not add additional nanoparticles to account for the partial volumes of the nanoparticles that are located outside of the cuboid. For both additional procedures, the overall number of nanoparticles consequently remains fixed a value of $N_0 = f \frac{V_{\text{dist}}}{V_{\text{scat}}}$ (see Chap. 3). Moreover, in all procedures, I ensure that there is no overlap between the nanoparticles. To visualize the differences between procedures I and II, Fig. A.2a, b present the local volume fractions of several nanoparticle distributions that were generated using the two procedures over the spatial coordinate z. For both procedures, these figures include several exemplary particle distributions with different lengths (l_{dist}). An analogous analysis for the procedure used in the main text (procedure I) is presented

D. Werdehausen, *Nanocomposites as Next-Generation Optical Materials*, Springer Series in Materials Science 316, https://doi.org/10.1007/978-3-030-75684-0

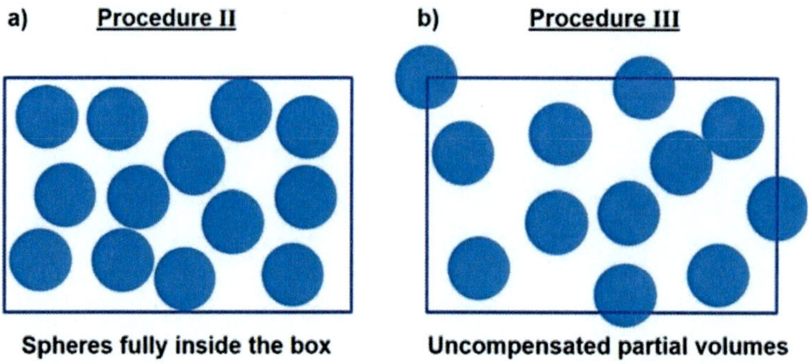

Fig. A.1 Alternative procedures for placing the scatterers. **a** The nanoparticles are forced to be completely inside the cuboid. **b** Only the center of the nanoparticles must be within the cuboid, but there is no compensation for the partial volumes that are located outside of the cuboid (δV_{out}). Reproduced from [1]

in Fig. A.3b. First, Fig. A.2a demonstrates that, if the nanoparticles are required to be fully within the cuboid (procedure II), the local volume fractions in the center region of the cuboid (around $z = 0$) are significantly higher than the target volume fraction of $f = 15\,\%$. In contrast, Fig. A.2a shows that, for procedure III, the average local volume fraction in this region is slightly below the target volume fraction (Fig. A.2b). It is evident from the figures that this difference is caused by the fact that a part of the nanoparticles' volumes can be located outside of the cuboid for procedure III.

To analyze the impact of the differences in the placement procedures on the effective refractive indices of the nanoparticle distributions, I performed an additional series of simulations. For purpose, I chose a fixed nanoparticle radius of $r_{scat} = 30\,\text{nm}$ and varied the volume fractions between $f = 5\,\%$ and $f = 25\,\%$. I kept all remaining parameters fixed as in the corresponding analysis in Chap. 3 ($\lambda_0 = 0.8\,\mu\text{m}$, $\omega_{beam} = \lambda_0$, $w_{dist} = 6\omega_{beam}$, and $l_{dist} = 20r_{scat}$). The effective refractive indices obtained from this analysis for procedure II and procedure II are presented in Fig. A.2c, d. First, Fig. A.2c demonstrates that, if the nanoparticles are completely inside the cuboid (procedure II), the effective refractive indices obtained from the transmitted component agree well with the EMT. Furthermore, they almost perfectly coincide with the results obtained for procedure I (see Chap. 4). In contrast, the effective refractive indices retrieved from the reflected component are significantly below the corresponding values obtained for procedure I and from the EMT. Finally, Fig. A.2d shows that, for procedure III, the effective refractive indices retrieved in both reflection and transmission are significantly lower than the values obtained for procedure I and from the EMT. This demonstrates that accounting for the partial volumes that are located outside of the cuboid is essential for obtaining well-defined volume fractions and reliable effective refractive indices.

In combination, the findings for all placement procedures show that the effective refractive indices obtained in transmission are mainly determined by the average

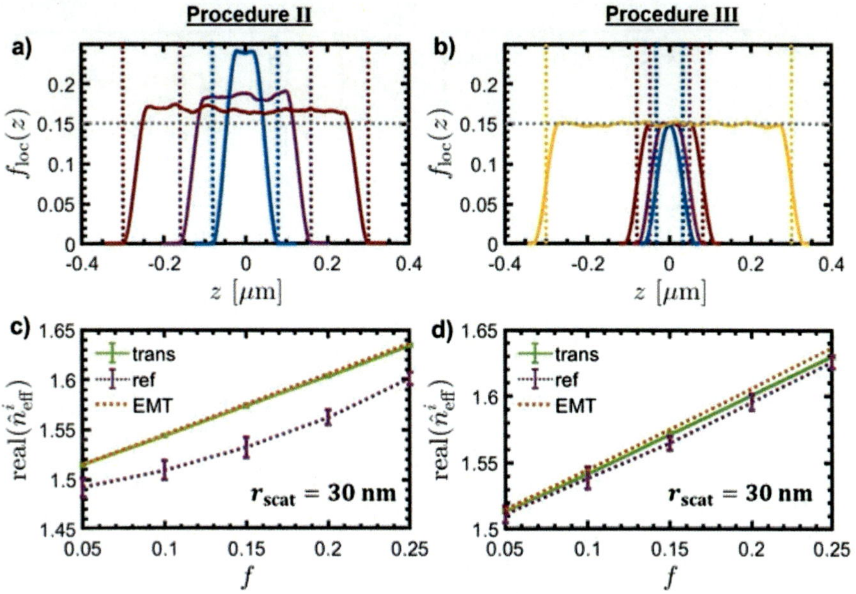

Fig. A.2 **a**, **b** Local volume fraction (f_{loc}) as a function of z for different values of l_{dist} for the two alternative placement procedures from Fig. A.1. The grey lines visualize the target volume fraction of $f = 15\,\%$. The nanoparticle radius was fixed at $r_{\mathrm{scat}} = 30$ nm. **c**, **d** Ensemble average of the real part of the effective refractive index (real($\hat{n}^i_{\mathrm{eff}}$) and imag($\hat{n}^i_{\mathrm{eff}}$)) as a function of the volume fraction (f) at $r_{\mathrm{scat}} = 30$ nm for the two alternative placement procedures. All plots include the data obtained from transmission, reflection, and the EMT ($i \in$ {trans, ref, EMT}). The ensemble averages were obtained by averaging across different microstates (random distributions) and the errorbars were determined as the corresponding standard deviations ($\Delta n^i_{\mathrm{eff}} = \sigma(\{n^i_{\mathrm{eff}}\})$. Reproduced from [1]

volume fraction within the cuboid. This follows from the finding that, for both procedure I and procedure II, for which the average volume fraction within the cuboid is exactly equal to the target value, the effective refractive indices agree well with each other and the EMT. In contrast, for procedure III, the overall volume fraction within the cuboid is slightly lower because there is no compensation for the partial volumes that are located outside the cuboid. This explains the slight decrease of the effective refractive index obtained in transmission that is observed in Fig. A.2d. Finally, the values obtained in reflection show that this component is very sensitive to the location of the first interface. This can be seen from the finding that, if the nanoparticles are forced to by fully inside the box (procedure II), the interfaces shift towards the center of the box (Fig. A.2a). Since this increases the distance the beam has to cover until it reaches the nanoparticle distribution, this explains the drastic decrease of the effective refractive indices obtained in reflection for procedure II (Fig. A.2c). Finally, this also demonstrates that the fact that there are no well-defined interfaces for larger scatterers is a key issue. As discussed in Sect. A.1.3 this can be compensated for by introducing an "effective length" or equivalently "effective interfaces".

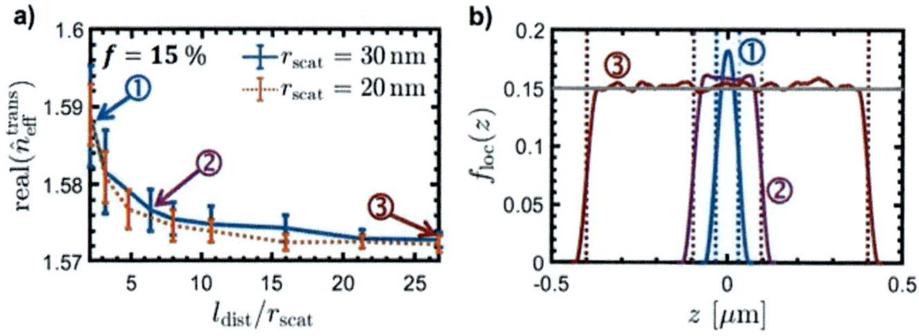

Fig. A.3 Convergence of the ensemble average of the real part of the effective refractive index retrieved from the transmitted component (real($\hat{n}_{eff}^{trans}$)) with increasing length ($\frac{l_{dist}}{r_{scat}}$) for nanoparticle radii of $r_{scat} = 20\,\mathrm{nm}$ and $r_{scat} = 30\,\mathrm{nm}$ ($f = 0.15\,\%$). The data was obtained by averaging over different random distributions. The error bars denote the corresponding standard deviations. For both radii, the width of the box was fixed at $w_{dist} = 6\omega_{beam} = 4.8\,\mu\mathrm{m}$ and the beam width at $\omega_{beam} = \lambda_0 = 0.8\,\mu\mathrm{m}$. Its evident that real($\hat{n}_{eff}^{trans}$) converges to a fixed value around $l_{dist} = 20r_{scat}$ for both radii. (b) Local volume fraction (f_{loc}) as a function of z along the box for three different values of l_{dist} and $r_{scat} = 30\,\mathrm{nm}$. Only for sufficiently large values of l_{dist} there is a well-defined volume fraction within the bulk. Reproduced from [1]

A.1.2 Convergence with Increasing Lengths

In this section, I analyze the convergence of the effective refractive index with increasing length (l_{dist}) for two additional nanoparticle radii to demonstrate that convergence is generally achieved for lengths of $l_{dist} = 20r_{scat}$. Moreover, I discuss that this length is required to achieve a well-defined volume fraction within the center, i.e. the bulk region, of the nanoparticle distributions.

First, Fig. A.3a shows that, for $r_{scat} = 20\,\mathrm{nm}$ and $r_{scat} = 30\,\mathrm{nm}$, the effective refractive index also converges to a fixed value at a length of $l_{dist} = 20r_{scat}$. Second, Fig. A.3b depicts the local volume fraction (f_{loc}) as a function of the spatial coordinate z for three nanoparticle distributions at different values of l_{dist}. For this analysis, the nanoparticles' radii and volume fractions remained fixed at $r_{scat} = 30\,\mathrm{nm}$ and $f = 15\,\%$, respectively. The data in Fig. A.3b shows that, for small values of l_{dist}, the local volume fraction displays a distinct peak at $z = 0$. Furthermore, the local volume fraction within the center of the particle distribution is only equal to the target value ($f = 0.15\,\%$ in Fig. A.3) for sufficiently large values of l_{dist}.

A.1.3 Effective Interfaces

Another key limitation for materials outside the homogeneous regime is that their interfaces are no longer well-defined. To illustrate this directly, Fig. A.4a presents the local volume fraction (f_{loc}) over the spatial coordinate z for different nanopar-

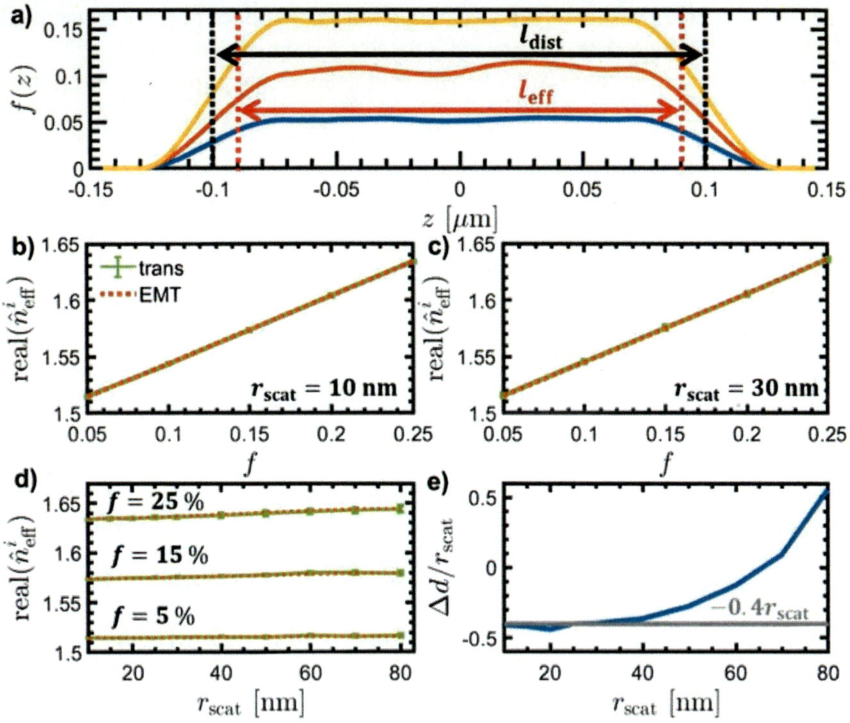

Fig. A.4 **a** Local volume fraction (f_{loc}) as a function of the spatial coordinate z to show that the interfaces are not well-defined for larger particles (here $r_{scat} = 30$ nm). Therefore, the effective length l_{eff} is determined by optimizing the length for the best agreement of real($\hat{n}^{trans}_{eff}$) with the EMT-prediction simultaneously for all volume fractions at each radius (r_{scat}). **b, c** Ensemble averages for the real part of the effective refractive index (real($\hat{n}^{i}_{eff}$)) using effective propagation lengths for (b) $r_{scat} = 10$ nm and (**c**) $r_{scat} = 30$ nm. **d** Analogous plot as a function of the particle radius (r_{scat}) at a volume fractions of $f = 5\,\%$, $f = 15\,\%$ and $f = 25\,\%$. The effective interface was determined for each radius individually. The data for the imaginary part is depicted in Fig. A.5. **e** Difference $\Delta l = (l_{dist} - l_{eff})$ between the length of the particle distribution (l_{dist}) and the effective length (l_{eff}) as a fraction of the radius (r_{scat}). Reproduced from [1]

ticle distributions with a radius of $r_{scat} = 30$ nm. This figure shows that the local volume fraction increases gradually to its bulk value across a region whose size is equal to the nanoparticles' size ($d_{scat} = 2r_{scat} = 60$ nm). While this is also the case for conventional materials, this effect does not play a role for such materials because their building blocks are atoms or small molecules, which are orders of magnitude smaller than the wavelength. However, all expressions that I use to retrieve the effective refractive indices from my numerical simulations depend directly on the length of the nanoparticle distributions (l_{dist}; see (3.1)). Since this quantity is no longer well-defined outside the homogeneous regime, the question arises if this can compensate for the finding that the effective medium theory (EMT) systematically overestimates the real part of the effective refractive index at larger volume fractions (see Fig. 3.5). To investigate whether this is true, I introduced an effective length (l_{eff}) as a free parameter and optimized it for the best agreement of real($\hat{n}_{\mathrm{eff}}^{\mathrm{trans}}$) with the EMT's prediction. Since an effective length should not depend on the volume fraction, I performed this optimization for all volume fractions ($5\,\% \leq f \leq 25\,\%$) simultaneously. First, I did so for two fixed nanoparticle radii of $r_{\mathrm{scat}} = 10\,\mathrm{nm}$ and $r_{\mathrm{scat}} = 30\,\mathrm{nm}$. For both nanoparticle radii, this led me to an effective length that is $0.4r_{\mathrm{scat}}$ shorter than l_{dist}. As shown in Fig. A.4b, c, using this length in the retrieval procedure yields a perfect agreement of the numerical results with the EMT. In addition, Fig. A.5 demonstrates that the level of agreement between the imaginary part of the effective refractive index (imag($\hat{n}_{\mathrm{eff}}^{\mathrm{trans}}$)) and the EMT remains unaffected for both nanoparticle radii (see Fig. 3.6 in Chap. 3) even though I did not include the imaginary part in my procedure for determining the effective length. Furthermore, Fig. A.4d shows that the concept of an effective propagation length also leads to an essentially perfect agreement between real($\hat{n}_{\mathrm{eff}}^{\mathrm{trans}}$) and the EMT for all other radii between $r_{\mathrm{scat}} = 10\,\mathrm{nm}$ and $r_{\mathrm{scat}} = 80\,\mathrm{nm}$. To investigate the influence of the nanoparticle radii on the effective length, Fig. A.4e presents the difference between l_{dist} and l_{eff} ($\Delta l = l_{\mathrm{dist}} - l_{\mathrm{eff}}$) over particle radius r_{inc}. These data show that for nanoparticle radii below $r_{\mathrm{scat}} = 40\,\mathrm{nm}$, the effective propagation length remains $0.4r_{\mathrm{scat}}$ shorter than l_{dist}, whereas a continuous increase is observed for larger particles. However, Δl_{eff} never exceeds the nanoparticles' radius (r_{scat}). The strong increase observed for larger nanoparticle radii possibly indicates that an additional physical mechanism starts to affect the effective propagation length in this regime. For example, it appears possible that the effective propagation length can account for the influence of the distributions' width that plays an increasing role for larger particles (see Sect. A.1.6). However, even for the large nanoparticle radii, my concept of an effective propagation length yields a perfect match between the numerical results and the EMT's prediction. Finally, the fact that changing the length used in the retrieval procedure by a fraction of r_{scat} highlights that the lack of well-defined interfaces is a key issue outside the homogeneous regime. In this regime, it is consequently not possible to assign an effective refractive index to a well-defined region in space.

A.1.4 Retrieval Equation for Metamaterials

In the form given in the main text, the Fresnel equations are only valid if the materials' permeabilities are unity ($\mu_1 = \mu_{\mathrm{eff}} = 1$). For scatterers, whose magnetic dipole response plays a role, this is no longer the case [2, 3]. Therefore, I here use an additional expression that relies on both r and t to determine n_{eff}. This equation is valid even for non-unity permeabilities [4, 5]:

$$\cos(n_{\mathrm{eff}}^{\mathrm{meta}} k l_{\mathrm{prop}}) = 2\frac{1 - r^2 + n_{\mathrm{h}} t^2}{[n_{\mathrm{h}}(1 + r) + 1 - r]t}. \tag{A.1}$$

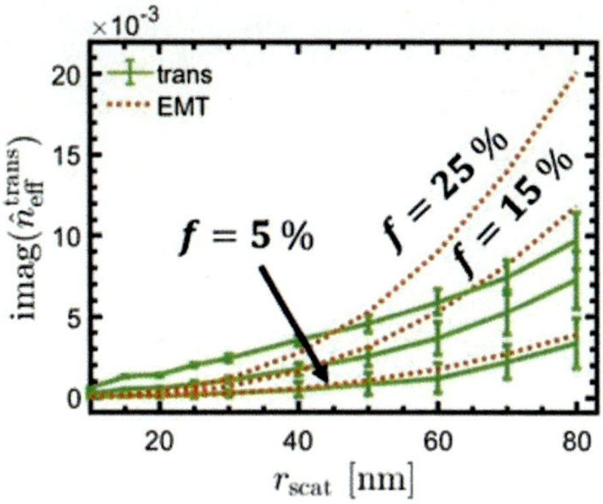

Fig. A.5 Under the assumption of a effective length (l_{eff}): Ensemble average of the imaginary part of the effective refractive index obtained in transmission (imag($\hat{n}_{\text{eff}}^{\text{trans}}$)) as a function of the nanoparticle radius (r_{scat}) at volume fractions of $f = 5\,\%$, $f = 15\,\%$, and $f = 25\,\%$. The imaginary part remains largely unaffected by the introduction of an effective length. Reproduced from [1]

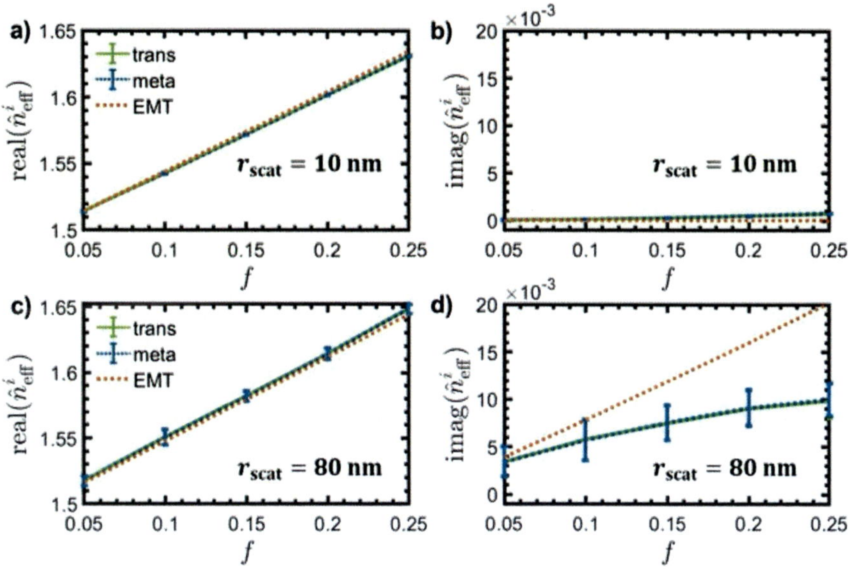

Fig. A.6 Ensemble average of the real (**a**, **c**) and imaginary (**b**, **d**) parts of the effective refractive index (real($\hat{n}_{\text{eff}}^{i}$) and imag($\hat{n}_{\text{eff}}^{i}$)) as a function of the volume fraction (f) for radii of at $r_{\text{scat}} = 10\,\text{nm}$ and $r_{\text{scat}} = 80\,\text{nm}$. All plots include the data obtained in transmission, from the equation for metamaterials, and from the EMT ($i \in \{\text{trans, meta, EMT}\}$). The finding that $n_{\text{eff}}^{\text{meta}}$ agrees almost perfectly with $n_{\text{eff}}^{\text{trans}}$ shows that the magnetic dipole response remains negligible for both radii. Reproduced from [1]

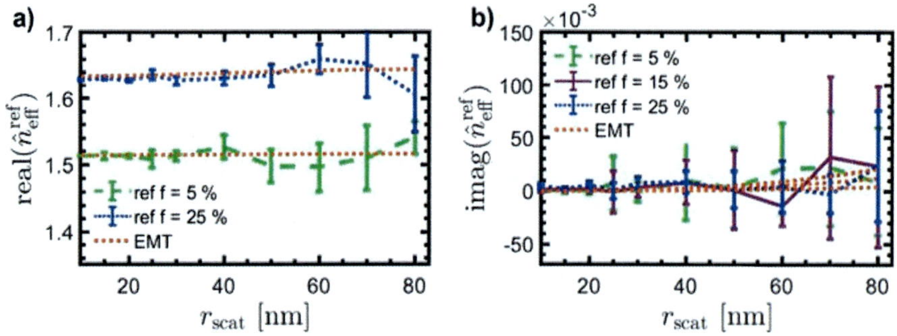

Fig. A.7 Ensemble average of (**a**) the real and (**b**) the imaginary parts of the effective refractive index in reflection (real($\hat{n}_{\rm eff}^{\rm ref}$) and imag($\hat{n}_{\rm eff}^{\rm ref}$)) as a function of the nanoparticle radius ($r_{\rm scat}$) at volume fractions of $f = 5\,\%$, $f = 15\,\%$, and $f = 25\,\%$. These data are not included in the main text to improve clarity. The dashed red lines visualize the results from the effective medium theory (EMT). Reproduced from [1]

Since this equation was originally derived for photonic-crystal-type materials [4, 5], I refer to the values obtained from this expression as $n_{\rm eff}^{\rm meta}$. Moreover, I eliminate the ambiguity that arises from inverting the cosine in (A.1) by requiring real($n_{\rm h}$) $\leq$ real($n_{\rm eff}$) $\leq$ real($n_{\rm scat}$).

To investigate if the values obtained from (A.1) deviate from those discussed in the main text, I here investigate the effective refractive indices of nanoparticle distributions with radii of $r_{\rm scat} = 10\,$nm and $r_{\rm scat} = 80\,$nm. These radii correspond to the smallest and largest radii that I included in my main analysis in Chap. 4. Accordingly, Fig. A.6 depicts the ensemble averages of the real and imaginary parts of the effective refractive indices obtained in transmission ($n_{\rm eff}^{\rm trans}$) and from the equation for metamaterials for both radii as a function of the volume fraction. It is evident that $n_{\rm eff}^{\rm meta}$ matches $n_{\rm eff}^{\rm trans}$ almost perfectly for both radii. This can be understood from the fact that the transmitted component highly dominates over the reflected one ($|t| \gg |r|$). Furthermore, this shows that the magnetic dipole response remains negligible even for the largest radii analyzed in the main text.

A.1.5 Ensemble Averages in Reflection

In this section, I present and discuss the data for the ensemble averages obtained in reflection ($\hat{n}_{\rm eff}^{\rm ref}$), which I did not include in Chap. 4 because they fluctuate heavily.

Figure A.7 demonstrates that for both the real (Fig. A.7a) and imaginary (Fig. A.7b) parts, the ensemble avarages retrieved from the reflected component only agree well with the EMT for small nanoparticle radii. For larger radii ($r_{\rm scat} > 30\,$nm) heavy fluctuations are present at all volume fractions. In the imaginary part, these fluctuations can extend over more than two orders of magnitude (Fig. A.7b). This

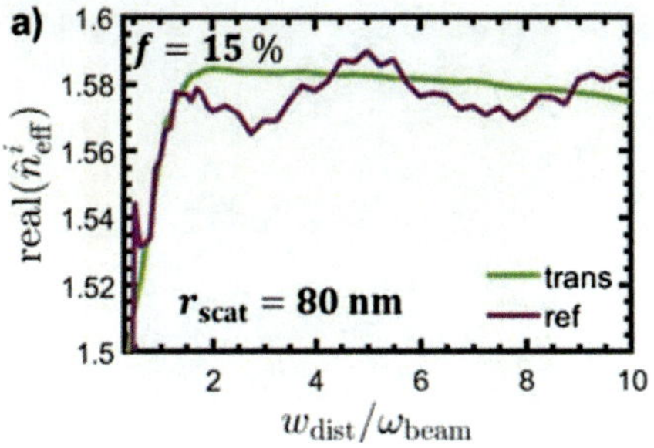

Fig. A.8 **a** Real and **b** imaginary parts of the effective refractive index in transmission ($n_{\text{eff}}^{\text{trans}}$) with increasing width ($\frac{w_{\text{dist}}}{\omega_{\text{beam}}}$) of the cuboid that encompasses the nanoparticles' centers for a radius of $r_{\text{scat}} = 80\,\text{nm}$. The data were obtained by first generating the particle distribution with the maximum width ($w_{\text{dist}} = 3\omega_{\text{beam}}$) and then cropping this distribution. Incoherent scattering causes the effective refractive index to depend on the width even for wide particle distributions. Reproduced from [1]

illustrates that the concept of an effective refractive index is not suitable to predict the reflection from a distribution of scatterers for $r_{\text{scat}} > 30\,\text{nm}$.

A.1.6 Width Dependence—The Influence of Incoherent Scattering

As discussed in Chap. 4, incoherent scattering plays an increasing role for larger nanoparticle radii. In this section, I demonstrate that this also causes the effective refractive indices to depend on the widths of the nanoparticle distributions (w_{dist}). For this analysis, I selected the maximum nanoparticle radius analyzed in Chap. 4 ($r_{\text{scat}} = 80\,\text{nm}$), because the amount of incoherent scattering increases with increasing radii.

Figure A.8 demonstrates that, for large values of r_{scat}, the effective refractive indices in both transmission and reflection depend strongly on the width of the nanoparticle distribution (w_{dist}). This holds true for both the real (Fig. A.8a) and imaginary (Fig. A.8) parts, even if the distribution is a factor of 10 wider than the beam. I again obtained these data by first generating the particle distribution with the maximum width ($w_{\text{dist}} = 10\omega_{\text{beam}}$) and then cropping this distribution to smaller widths. As discussed in Chap. 4, the width dependence can be attributed to incoherent scattering. This is because light that is scattered out of the forward and backward directions leads to the emergence of long-ranged modes that propagate along the entire width of the distribution and hence affect its properties. Since, for homogeneous materials, the effective refractive index is independent of the width (see main text), this adds an additional level on which the concept of an effective refractive index breaks down outside the homogeneous regime.

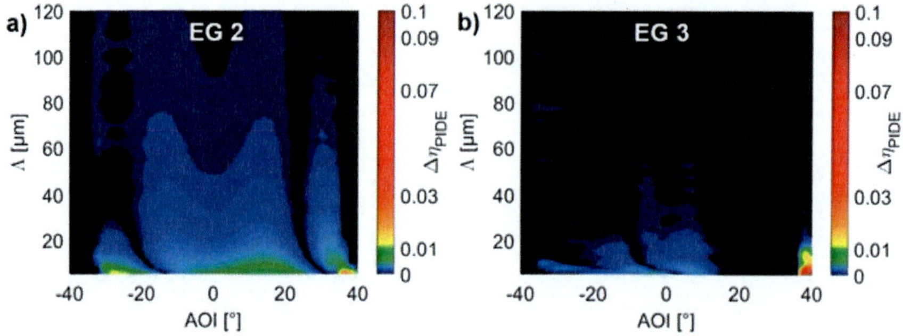

Fig. A.9 Difference in the polychromatic integral diffraction efficiency between TM and TE polarization ($\Delta\eta_{\mathrm{PIDE}} = |\eta^{\mathrm{TM}}_{\mathrm{PIDE}} - \eta^{\mathrm{TE}}_{\mathrm{PIDE}}|$) for EG 2 and EG 3 (see Table 5.1) over the grating period (Λ) and the angle of incidence (AOI). It is evident that $\Delta\eta_{\mathrm{PIDE}}$ only exceeds 1 percentage point for grating periods of $\Lambda \leq 10\ \mu\mathrm{m}$. Reprinted with permission from [6] © The Optical Society

A.2 Polarization Dependence of Nanocomposite-Enabled DOEs

To demonstrate that the nanocomposite-enabled EGs' performance is almost independent of the polarization, Fig. A.9 depicts the difference in the η_{PIDE} between TM and TE polarization ($\Delta\eta_{\mathrm{PIDE}} = |\eta^{\mathrm{TM}}_{\mathrm{PIDE}} - \eta^{\mathrm{TE}}_{\mathrm{PIDE}}|$) for the nanocomposite-enabled EGs EG 2 and EG 3 (see Table 5.1). The figure shows that the efficiency difference between the two orthogonal polarizations stays below 1 percentage point across almost all grating periods and AOIs included in the simulations. Slightly larger differences are only observed for $\Lambda < 10\ \mu\mathrm{m}$. This shows that nanocomposite-enabled EGs are indeed suitable for applications that operate with unpolarized light.

A.3 Performance Benefits of Nanocomposite-Enabled DOEs Over the State-of-the-Art Solutions

In this section, I show that nanocomposite-enabled EGs can indeed provide unprecedented high efficiencies. To this end, I use the same methods as in Chap. 5 to briefly analyze the state-of-the-art solutions for achieving achromatic phase and efficiency profiles. A comprehensive overview over these state-of-the-art solutions is given in [7].

The best known concept for achieving achromatic efficiency profiles has been dubbed "multilayer EG" [7–12]. This approach also builds on a two-layer structure, but allows the two layers to take different heights (see visualization in the inset of Fig. A.10a). Within the TEA, the maximum phase delay per period for such a structure consequently reads:

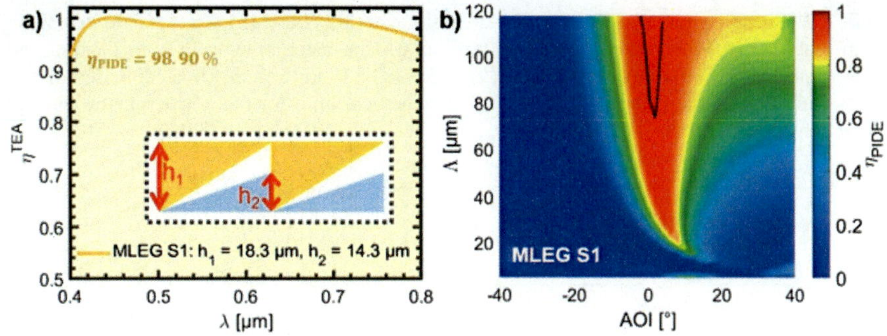

Fig. A.10 **a** Efficiency of the state-of-the-art multilayer EG (MLEG S1; specifications in Table A.1) from the TEA over the wavelength. The inset illustrates that a multilayer EG is generally made of two layers, which have different heights (h_1 and h_2). **b** η_{PIDE} of MLEG S1 from full wave optical FEM simulations (JCMsuite) as a function of the the grating period (Λ) and the angle of incidence (AOI). The dotted black line highlights the broad region across which η_{PIDE} stays within 3 % of its theoretical limit. Adapted with permission from [13] © The Optical Society

$$\phi_{\max}^{\text{TEA}}(\lambda) = \frac{2\pi h_1}{\lambda}\left[n_1(\lambda) - 1\right] + \frac{2\pi h_2}{\lambda}\left[n_2(\lambda) - 1\right], \qquad \text{(A.2)}$$

where $n_1(\lambda)/n_2(\lambda)$ and h_1/h_2 are the refractive indices and the heights of the two layers, respectively. Equation (A.2) directly shows that a multilayer EG has two degrees of freedom (the heights h_1 and h_2), which can be adjusted to achieve $\phi_{\max}^{\text{TEA}}(\lambda_1) = 2\pi = \phi_{\max}^{\text{TEA}}(\lambda_2)$ and hence an efficiency of 100 % for two distinct design wavelengths (λ_1 and λ_2). This is possible for any two materials that have different refractive indices. As a typical example for a multilayer EG, I here analyze a structure whose two layers are made of PMMA and PC [7]. For the design of this structure, I again optimized the design wavelengths (the heights) such that its η_{PIDE} is maximized. All the EG's specifications resulting from this optimization are listed in Table A.1. Moreover, Fig. A.10a depicts this EG's efficiency (η^{TEA}) over the wavelength. This figure demonstrates that the multilayer approach also allows for achieving $\eta_{\text{PIDE}}^{\text{TEA}} \approx 99\,\%$. However, the results of the FEM simulations for MLEG S1 in Fig. A.10b show that, within the entire range of grating periods and AOIs included in this figure, this EG's η_{PIDE} remains significantly below its theoretical limit ($\eta_{\text{PIDE}}^{\text{TEA}} \approx 98.90\,\%$). In addition, its η_{PIDE} drops rapidly below 90 % for increasing AOIs (particularly for negative AOIs) and decreasing grating periods. This is probably one of the key reasons why these multilayer EGs, which have been already integrated into a commercial camera lens [8], are not used in a wider range of optical systems. In fact, for this structure, it is not possible to keep the amount of stray light in spurious diffraction order below 1 %, which is a common requirement for high-end optical systems (see Sect. 5.2.4). Furthermore, the narrow region within which high values of η_{PIDE} are maintained leads to harsh constraints on the optical design.

The second state-of-the-art approach that has been discussed in the literature [7] is the use of the same two-layer approach as for nanocomposite-enabled EGs (Fig. 5.3),

Table A.1 Specifications of the state-of-the-art multilayer (ML) EG (MLEG S1) and the "common depth" (CD) EG (EG S2). Materials 1 and 2 correspond to the materials in the EGs' first and second layer, respectively. The sum for the height (h) of MLEG S1 indicates that it is composed of two layers with different heights. $\eta_{\mathrm{PIDE}}^{\mathrm{TEA}}$ denotes the theoretical limit for the polychromatic integral diffraction efficiency that is obtained from the thin element approximation (TEA).

Name	Type	Material 1	Material 2	h	$\eta_{\mathrm{PIDE}}^{\mathrm{TEA}}$
MLEG S1	ML	PMMA	PC	18.3 μm + 14.3 μm	98.90 %
EG S2	CD	N-BaF52	PC	26.3 μm	96.91 %

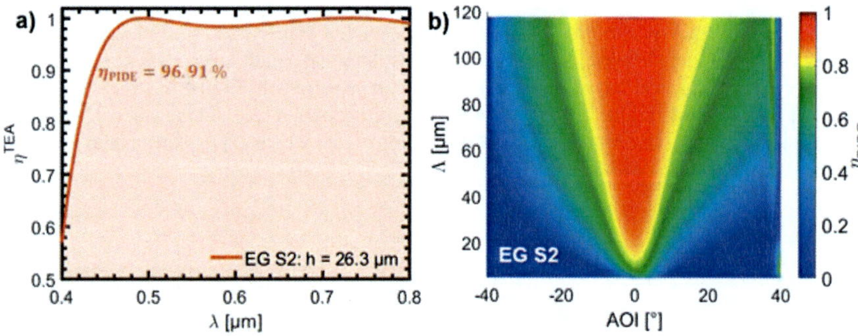

Fig. A.11 **a** Efficiency of the state-of-the-art common depth EG (EG S2; specifications in Table A.1) from the TEA over the wavelength. **b** η_{PIDE} of EG S2 from full wave optical FEM simulations (JCMsuite) as a function of the the grating period (Λ) and the angle of incidence (AOI). No dotted line is included in the figure since the maximum η_{PIDE} within entire range is only 93.88 %, which is not within 3 % of this EG's theoretical limit ($\eta_{\mathrm{PIDE}}^{\mathrm{TEA}} = 96.91$ %). Reprinted with permission from [13] © The Optical Society

but with conventional materials instead of dispersion-engineered nanocomposites. This concept has been dubbed "common depth EG" in [7]. But, because even a single-layer EG is a common depth EG whose second material is air, I here simply refer to these structures as EGs. The most discussed material combination for which relatively high efficiencies can be achieved is an EG whose first and second layers are made of the optical glass N-BaF52 and the polymer PC, respectively [7, 12, 14]. This EG is denoted as EG S2 in Table A.1. Figure A.11a shows that the theoretical limit for the broadband efficiency is only $\eta_{\mathrm{PIDE}}^{\mathrm{TEA}} = 96.91$ % for this EG's material combination. Furthermore, the data from the FEM simulations for this device (Fig. A.11b) illustrate that this EG's η_{PIDE} drops rapidly with decreasing grating periods and increasing AOIs. In fact, the lack of a dotted black line in this figure highlights that this EG's η_{PIDE} doesn't get within 3 % of its theoretical limit within the entire figure. Specifically, the maximum η_{PIDE} within the figure, which is reached for normal incidence and the largest grating period of $\Lambda = 120$ μm, is only 93.88 %. This shows that the dispersion engineering capabilities of nanocomposites are the key to achieving average efficiencies of almost 100 % and a high performance across a wide range of grating periods and AOIs. I discuss this aspect in detail in Sect. 5.3.1, where I develop a general design formalism for broadband DOEs.

A.4 Achromatic Metalenses

Recently the notion that achromats must consist of two optical elements was challenged by the demonstration of "achromatic metalenses" [15–18]. If this technology fulfilled the requirements of optical systems, it could have a major impact. Therefore, I here analyze this approach from the DOE design perspective. To do so, I first reconcile the principle behind these devices with the description of diffractive lenses I use throughout this book (see (2.19)). Based on this common description I then discuss their suitability for optical systems.

From the equation for the wavelength dependence of a DL ($f(\lambda) = \frac{\lambda_0}{\lambda} f(\lambda_0)$; see (2.20), it directly follows that achromatization ($f(\lambda_1) = f(\lambda_0)$) can be achieved by scaling the expression for $f(\lambda_1)$ by an additional factor $c_{\text{acr}}(\lambda_1)$. Doing so leads me to $f(\lambda_1) = c_{\text{acr}}(\lambda_1)\frac{\lambda_0}{\lambda_1} f(\lambda_0) = f(\lambda_0)$, and hence to $c_{\text{acr}}(\lambda_1) = \frac{\lambda_1}{\lambda_0}$. The phase profile of an achromatic diffractive singlet consequently reads:

$$\phi_{\text{acr}}(r, \lambda_1) = -\frac{\pi}{\lambda_0 c_{\text{acr}}(\lambda_1) f(\lambda_0)} r^2 = -\frac{\pi}{\lambda_1 f(\lambda_0)} r^2. \tag{A.3}$$

This simple expression demonstrates that precisely tailoring the phase profile as a function of both the radius and the wavelength theoretically allows for the design of a perfectly achromatic singlet. In other words, such a device would be a single DL which has the same focal length for all wavelengths. Furthermore, (A.3) shows that it is not necessary to use a Taylor expansion that it restricted to a limited spectral range and necessitates unnecessarily complicated language like "relative group delay" and "group delay dispersion" as was used in [15]. To visualize this relationship, the dotted lines in Fig. A.12 exemplarily depict the phase profiles of an achromatic singlet for three wavelengths (λ_g, λ_d, and λ_C) and a total phase delay of $\phi_{\max}(\lambda_d) = 10\pi$. It is evident that the overall curvature and hence phase difference across the element decreases with increasing wavelength. However, the phase differences that can be achieved in metalenses with manufacturable heights are highly restricted. Therefore, it is again necessary to remove unnecessary multiples of 2π from the continuous phase profiles in (A.3). The solid lines in Fig. A.12 show that achromatic focusing can then be achieved by varying the local widths of the different segments as a function of the wavelength. In fact, repeating the analysis that lead to (5.3) for the phase profile given in (A.3), yields a simple expression for $\Lambda(r, \lambda_1)$:

$$\Lambda(r, \lambda_1) = \frac{1}{k_{\text{seg}}(r, \lambda_1)} = \frac{1}{\frac{\mathrm{d}}{\mathrm{d}r} N(r, \lambda_1)} = \frac{\lambda_1 |f(\lambda_0)|}{r} = \frac{\lambda_1}{\lambda_0} \Lambda(r, \lambda_0). \tag{A.4}$$

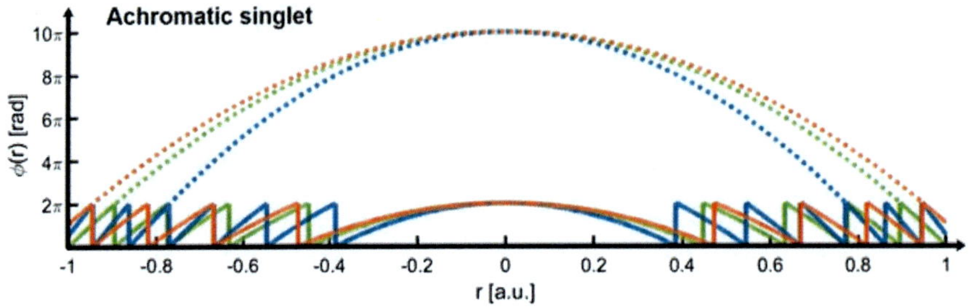

Fig. A.12 Phase profiles of an achromatic diffractive singlet at three different wavelengths (A.3). The different colors correspond to λ_g (blue), λ_d (green), and λ_C (red). After subtracting the unnecessary multiples of 2π, the local segment width (Λ_{seg}) must increase with the wavelength

This relationship shows that perfect achromacy can be achieved by linearly varying the local segment width as a function of the wavelength. This finding can also be readily confirmed using the grating equation, which for perpendicular incidence reads $\sin(\theta_q(\lambda_0)) = \frac{q\lambda_0}{\Lambda(\lambda_0)}$ (see (2.15)), where m is the integer that denotes the diffraction order. Specifically, starting from this expression and requiring a wavelength-independent diffraction angle ($\theta_q(\lambda_1) = \theta_q(\lambda_0)$ again yields (A.4).

The fact that the grating periods must be varied as a function of the wavelength directly leads me to the major challenge in the practical realization of achromatic metasurfaces: As discussed before, efficiencies of close to 100 % can only be achieved if each segment has a linear phase profile with a total phase difference that is equal to 2π for all wavelengths. In particular, it is essential to realize sharp phase jumps at the segments' edges for each wavelength. Figure A.12 visualizes that fulfilling this condition across the entire diameter of the lens entails realizing sharp phase jumps for one specific wavelength, while still continuously changing the phase delays for all other wavelengths. Fulfilling this condition with an accuracy that is high enough to achieve efficiencies of close to 100 % for all wavelength is a major challenge that is most likely impossible to overcome. This is also confirmed by the generally low and wavelength-dependent efficiencies reported in various publications (see overview in [19]). In particular, it is important to keep in mind that the ideal phase profiles can only be discretely sampled in metalenses [15]. Therefore, it is impossible to continuously vary the phase delays and sharp jumps can also not be realized for each wavelength individually. Furthermore, as opposed to the devices presented in [15, 18], a device that is suitable for conventional imaging system must also be polarization-independent. Finally, the fact that the ideal phase profiles are only discretely sampled also leads to a secondary color error (focal shift) that is essentially impossible to correct. This is because, it is not possible to implement a different local segment width (Λ_{seg}) for each wavelength individually with only a finite number of discrete phase steps. Therefore, all wavelengths between the finite number of design wavelengths experience the same grating period. These wavelengths hence undergo the dispersion of a conventional DL ($\nu_d = 3.452$). Between the design wavelengths, there is consequently a large remaining color error. In fact, this focal shift exhibits

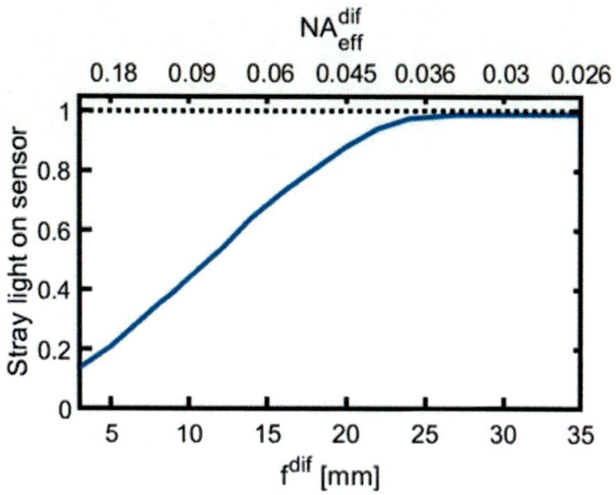

Fig. A.13 Percentage of the light in the second order that reaches a sensor with a diameter of $D_{sens} = 0.4$ mm which is placed into the focus of the first order of a hybrid two-element system. The refractive lens' focal length was fixed at $f^{ref} = 5$ mm and its aperture at $D = 2$ mm. The diffractive lens was modeled as an infinitely thin phase plate. At low values of NA^{dif} (large f^{dif}) all light in the second order reaches the sensor and hence reduces the contrast

sharp jumps if the neighboring design wavelength is reached. This discontinuity in the focal length is impossible to correct with additional (conventional) optical elements in an optical system. This effect is also directly evident from the data presented in the supplementary material of [20].

A.5 Stray Light in Hybrid Systems

For weakly focusing DLs in hybrid systems, the different orders propagate so close to each other that most of the light in the spurious orders reaches the sensor. In contrast, for high values of NA^{dif}, the light in the spurious diffraction quickly spreads out and is hence not visible on the sensor. To visualize and quantify this relationship, I used the wave propagation method (WPM) [21] to determine the amount of stray light that reaches the sensor in a hybrid two-element system with different values of NA^{dif}. For these simulations, I fixed the focal length of the refractive lens at $f^{ref} = 5$ mm, the aperture at $D = 2$ mm, and the wavelength at $\lambda_d = 0.588$ nm. Furthermore, I treated the DL as an ideal infinitely thin phase plate (see (2.19)) and performed a simulation for both the first and the second order. I then placed a sensor with a diameter of $D_{sens} = 0.4$ mm into the focus of the first order and determined the percentage of the second order's overall power flux that reaches the sensor. Accordingly, Fig. A.13 shows that, for low values of NA$^{dif}_{eff}$ (large f^{dif}), essentially all stray light in the second order hits the sensor. Since, as shown in Sect. 5.2.4, DLs in hybrid systems must be

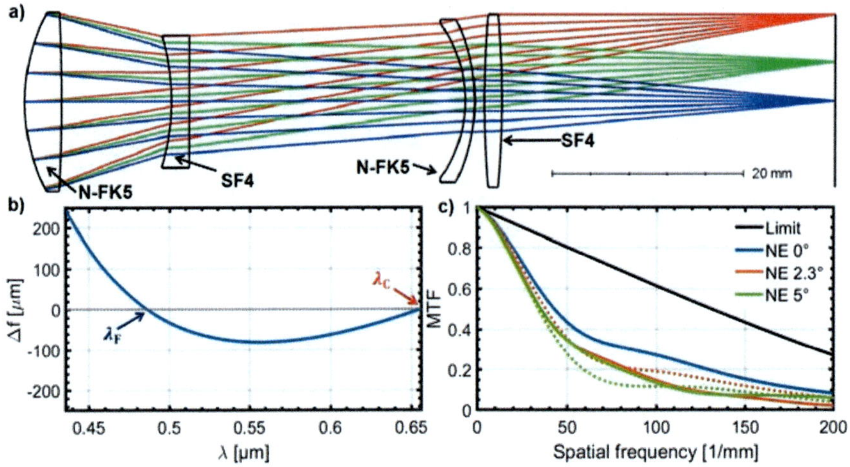

Fig. A.14 **a** Layout of a prototypical telephoto lens composed of four refractive elements. The system has a focal length of $f = 100\,\text{mm}$, a field of view of FOV $= 10°$, a f-number of $N_f = 5.6$, and a total system length of TL $= 85\,\text{mm}$. **b** Focal shift Δf over the wavelength (λ). Its evident that the system only fulfills the condition $f(\lambda_F) = f(\lambda_C)$. **c** Modulation transfer function (MTF) for different AOIs in both the sagittal (dashed lines) and tangential planes (solid lines). It is evident that the system's performance is drastically decreased compared to the system that includes a diffractive lens (DL; see Fig. 5.10)

operated at low values of $\text{NA}_{\text{eff}}^{\text{dif}}$ to achieve achromatization, this shows that most of the light in spurious diffraction order in such systems indeed reaches the sensor. Therefore, the contrast requirement of $\frac{1}{100}$ of high-end optical systems can only be fulfilled if the focusing efficiency remains within around 1 % of the theoretical limit.

A.6 Prototype Telephoto Lens—Refractive Benchmarks

To illustrate that diffractive lenses (DLs) are very powerful tools for correcting chromatic aberrations, I here present two benchmark designs that are only comprised of refractive elements for the prototype telephoto lens from Sect. 5.2.4. To this end, I keep the specifications and the basic configuration of the system fixed (see Fig. 5.10). In particular, I keep the constraint that the second lens group (the last two lenses) must be made of the same materials as the first group to ensure comparability. For the first design, I then simply left out the DL, whereas for the second design I replaced the DL by a refractive element. Note that I optimized all designs using a global optimization scheme until no further improvement was achieved for 24 hours. This, with high certainty, ensures that the designs presented here correspond to the global optima.

For the first design, Fig. A.14 depicts the layout (Fig. A.14a), focal shift as a function of the wavelength (Fig. A.14b), and modulation transfer function (MTF) as a function of the spatial frequency. The comparison of these figures to those presented in Fig. 5.10 show that the system's performance is drastically impaired without the DL if the DL is realized as a nanocomposite-enabled DL. Specifically, Fig. A.14b demonstrates the system is not apochromatic, but rather only achromatic ($f(\lambda_F) = f(\lambda_C)$). This design example hence already demonstrates that the DL allows for a significantly improved performance. However, it is important to keep in mind that leaving out the DL reduces the degrees of freedom that are available in the optimization process. Furthermore, because of the restriction to only two different materials, achieving apochromatization is (almost) impossible for such a system [22]. Therefore, a more suitable benchmark is a system in which the DL is replaced by a refractive lens. Accordingly, Fig. A.15 depicts an analogous figure for a system that is composed of five refractive lenses. This system's MTF (Fig. A.15c) demonstrates that adding the additional refractive lens indeed increases the performance compared to the system with only four lenses (Fig. A.14). However, the performance is still drastically below that of the system that includes the DL (Fig. 5.10 in Sect. 5.2.4). In fact, the system with five refractive lenses is also not apochromatic (Fig. A.15b). This is because forcing the system into an aprochromatic state highly increases the monochromatic aberrations and hence reduces the system's performance. These design examples consequently demonstrate that DLs are very powerful tools for correcting chromatic aberrations in broadband imaging systems. Since I have shown in the main text that nanocomposite-enabled DOEs fulfill all requirements of such systems, this indicates that this approach could indeed be an enabling technology for a new generation of optical systems.

A.7 Efficiencies of GRIN DOEs

To show that a nanocomposite-enabled GRIN DOE exhibits essentially the same performance as an equivalent échelette-type grating (EG), Fig. A.16b depicts the polychromatic integral diffraction efficiency (η_{PIDE}) of the GRIN DOE (see Fig. 5.19a), which is made of the same materials as EG 2 (see Table 5.1), as a function of the grating period (Λ) and the angle of incidence (AOI). This figure demonstrates that the performance of the GRIN DOE is almost identical to that of the corresponding EG (see Fig. 5.4b in Sect. 5.1). Specifically, just like the EG, the GRIN DOE exhibits a very high performance across a broad area in the diagram. To visualize this, the dotted black line in Fig. A.16 again highlights the region within which η_{PIDE} remains within 3 % of its theoretical limit.

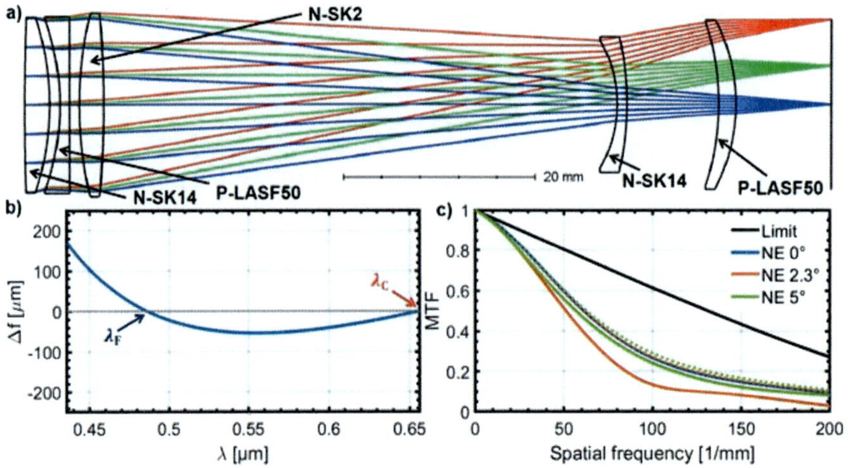

Fig. A.15 **a** Layout of a prototypical telephoto lens composed of five refractive elements. The system has a focal length of $f = 100\,\text{mm}$, a field of view of FOV $= 10°$, a f-number of $N_f = 5.6$, and a total system length of TL $= 85\,\text{mm}$. **b** Focal shift Δf over the wavelength (λ). Its evident that the system only fulfills the condition $f(\lambda_F) = f(\lambda_C)$. **c** Modulation transfer function (MTF) for different AOIs in both the sagittal (dashed lines) and tangential planes (solid lines). The fact that the system's performance is drastically decreased compared to the system that includes a diffractive lens (DL; see Fig. 5.10) illustrates that DLs are very powerful tools for correcting chromatic aberrations. Reproduced from [23]

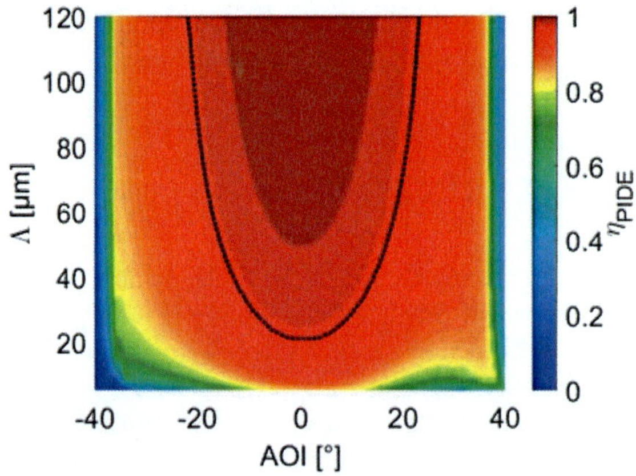

Fig. A.16 Polychromatic integral diffraction efficiency (η_{PIDE}) of a GRIN DOEs (see Fig. 5.19a), which is made of the same materials as EG 2 (see Table 5.1), as a function of the grating period (Λ) and the angle of incidence (AOI). The data was obtained from full wave optical FEM simulations (JCMsuite). The dotted black line indicates the broad region within which η_{PIDE} stays within 3 % of its theoretical limit. The theoretical limit corresponds to $\eta_{\text{PIDE}}^{\text{TEA}}$, which is the value that is obtained from the thin element approximation (TEA)

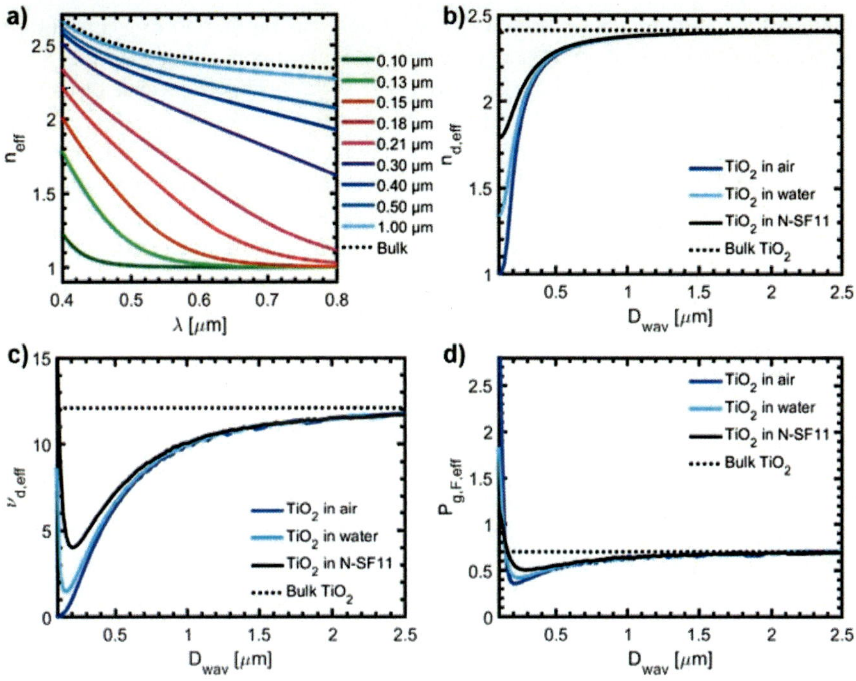

Fig. A.17 **a** Mode index (n_{eff}) of TiO_2 nanopillars in air at different diameters D_{wav}. **b** Mode index at the d-line ($n_{d,eff}$) as a function of (D_{wav}) for different cladding materials. **c**, **d** Analogous plots for the effective Abbe number ($\nu_{d,eff}$) and partial dispersion ($P_{g,F,eff}$). The dotted black lines in the figures denote the corresponding properties of bulk TiO_2. I chose water ($n_d = 1.33$, $\nu_d = 55.8$, and $P_{g,F} = 0.51$) and the glass N-SF11 ($n_d = 1.78$, $\nu_d = 25.8$, and $P_{g,F} = 0.61$) as the cladding materials to show influence of the cladding's refractive index

A.8 Towards Efficiency Achromatized Metalenses

As discussed in the main text (Sect. 5.4), the polychromatic integral diffraction efficiencies (η_{PIDE}) of current metalens designs are significantly below those achieved for single-layer EGs (see Table 5.1 and overview in [19]). In this section, I therefore analyze why efficiency-achromatization was not achieved for these state-of-the-art designs and suggest approaches for how this could become possible. I here focus on the waveguiding regime, because it allows to reduce the influence of shadowing effects and appears to be the most promising regime for the design of broadband metalenses (see Sect. 5.4 and [24]).

To focus on the fundamental physical effects, I here investigate the ideal case of uncoupled waveguides and hence investigate metagratings that are composed of independent waveguides. In this limit, the phase delay does not depend on the bulk refractive index but rather on the waveguides' mode index. For my analysis, I only consider the fundamental mode, since the waveguides must be operated in the single mode regime, for the phase delay to be well defined. Furthermore, I concentrate on waveguides made of TiO_2 and only briefly touch upon other materials since TiO_2 was used in most experimental works [15, 24–27]. This is because of its high transparency and refractive index in the visible spectral range. The latter is desirable because it allows for a strong confinement of the mode into the waveguide and thus reduces coupling between neighboring pillars. Finally, I only consider cylindrical waveguides, since they allow for the design of polarization insensitive metalenses [24, 26, 27]. Figure A.17a depicts the fundamental mode's effective refractive index (n_{eff}) of a cylindrical waveguide made of TiO_2 in air at different diameters (D_{wav}) as a function of the wavelength. I determined this data using the openly available Matlab toolbox introduced in [28]. This figure shows that, for large diameters, the mode indices converge to those of bulk TiO_2. However, already for diameters in the order of $D_{\mathrm{wav}} = 1\ \mu\mathrm{m}$, waveguide dispersion starts to play a major role. Further decreasing D_{wav} highly decreases the mode index while simultaneously increasing the overall dispersion. Finally, for very small radii, n_{eff} converges to the refractive index of the surrounding (cladding) material ($n_{\mathrm{clad}} \approx 1$ for air). To be able to connect these results to my design rules, I now characterize the mode index by its effective index at the d-line ($n_{\mathrm{d,eff}}$), Abbe number ($\nu_{\mathrm{d,eff}}$), and partial dispersion ($P_{\mathrm{g,F,eff}}$), which I define in analogy to the corresponding parameters of conventional materials (see (2.2)). Figure A.17b–d depict these parameters as functions of D_{wav} for different cladding (surrounding) materials. As additional cladding materials besides air, I also used water ($n_{\mathrm{d}} = 1.33$, $\nu_{\mathrm{d}} = 55.8$, and $P_{\mathrm{g,F}} = 0.51$) and the optical glass N-SF11 ($n_{\mathrm{d}} = 1.78$, $\nu_{\mathrm{d}} = 25.8$, and $P_{\mathrm{g,F}} = 0.61$) to investigate the influence of the magnitude of the cladding material's refractive index. First, Fig. A.17b confirms that the mode index starts at the refractive index of the core material for large values of D_{wav} and converges to that of the surrounding material for small values of D_{wav}. Second, Fig. A.17c shows that $\nu_{\mathrm{d,eff}}$ increases with decreasing D_{wav}. This shows that the overall amount of dispersion increases with decreasing waveguide diameters because of waveguide dispersion. Furthermore, comparing the curves for the different cladding materials demonstrates that, at small radii for which the mode is no longer completely confined in the core, the cladding material affects the mode's properties. Specifically, the smaller the refractive index difference between core and cladding, the earlier the cladding's refractive index starts to play a role. This shows that changing the cladding material is one potential approach for tailoring the waveguides dispersion.

To investigate whether tailoring the waveguide dispersion can, in principle, allow for the design of efficiency achromatizized metagratings, I now again use my design framework from Sect. 5.3.1. For this purpose, I used the refractive index data of TiO_2 for $n_1(\lambda)$ and vary only $n_{\mathrm{d},2}$ and $\nu_{\mathrm{d},2}$. Furthermore, I fixed $P_{\mathrm{g,F},2}$ at an intermediate value of $P_{\mathrm{g,F},2} = 0.51$ since I have already shown that the influence of the partial dispersion is relatively small. Accordingly, Fig. A.18a depicts η_{PIDE} as a function of $n_{\mathrm{d},2}$ and $\nu_{\mathrm{d},2}$. This figure again confirms that efficiency achromatization only occurs in a narrow region. To investigate whether the waveguides' effective parameters are located within this corridor, the dotted lines in Fig. A.18a present the behavior of the waveguides' $n_{\mathrm{d,eff}}$ and $\nu_{\mathrm{d,eff}}$ for the different cladding materials (from Fig. A.17). The black line shows that if the surrounding material is just air, $n_{\mathrm{d,eff}}$ and $\nu_{\mathrm{d,eff}}$

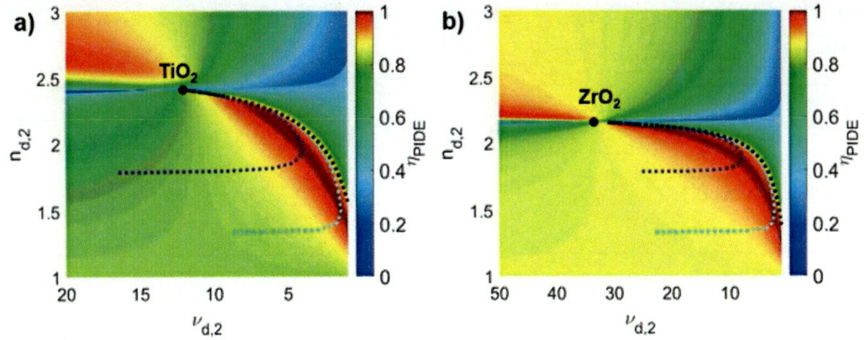

Fig. A.18 **a** Polychromatic integral diffraction efficiency (η_{PIDE}) as a function of $n_{\mathrm{d},2}$ and $\nu_{\mathrm{d},2}$, where the subscript 2 refers to $n_2(\lambda)$. For $n_1(\lambda)$ the refractive index profile of TiO_2 was used. The partial dispersion of $n_2(\lambda)$ was fixed at $P_{\mathrm{g,F},2} = 0.51$. The dotted lines visualize the waveguide dispersion as functions of D_{wav} in analogy to Fig. A.17 for different cladding materials (black: air, cyan: water, and blue: N-SF11). **b** Analogous illustration for ZrO_2. See Sect. 5.3.1 for details on how both (**a**) and (**b**) were obtained

lie just outside the region in which efficiency achromatization occurs. This shows that efficiency achromatization is not possible with a simple metalens composed of cylindrical TiO_2 pillars in air. This also explains why the experimentally measured efficiency curves of such devices (for example in [27]) show a strong wavelength dependence. To show that this finding is independent of the waveguides' material, Fig. A.18b depicts the results of an identical analysis for ZrO_2 instead of TiO_2. This figure shows that, independent of the material, the theoretical limit for the maximum η_{PIDE} of such designs is in the order of 90 % even under the assumption of an ideal linear phase profile. This demonstrates that more complicated approaches are needed to achieve efficiency achromatization for metalenses. In fact, Fig. A.18 demonstrates that it is necessary to slightly decrease the overall amount of dispersion (increase $\nu_{\mathrm{d,eff}}$). As discussed previously, one way of doing so would be to change the surrounding material by immersing the metalens in a liquid or adding coating. The cyan and blue lines in Fig. A.18 show that this can indeed shift $n_{\mathrm{d,eff}}$ and $\nu_{\mathrm{d,eff}}$ into the region in which efficiency achromatization occurs for some values of D_{wav}. However, this approach has a major disadvantage which is that decreasing the refractive index difference between cladding and core decreases the mode's confinement to core. Therefore, larger distances between the waveguides would be required to decouple their modes. This, in turn, would decrease the overall efficiency because of the much coarser sampling of the phase profiles [29]. Other approaches that could allow for the design of efficiency achromatized metagratings is the use of more complex waveguide shapes. In fact, this approach together with the geometric phase (Berry phase) was recently used to tailor the phase profile of a metasurface such that the longitudinal chromatic aberrations are reduced [15] (see Sect. A.4). In this case, the challenge is to find a design that is also polarization insensitive and can be readily fabricated. Finally, a very elegant possibility of tailoring the dispersion would be adjusting the

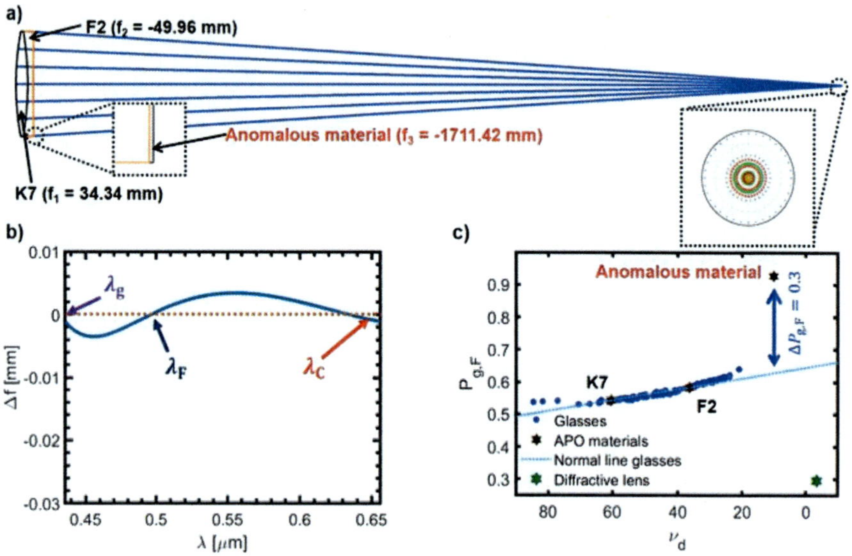

Fig. A.19 **a** Layout of an apochromat that is comprised of the optical glasses F2 and K7 as well as a thin layer of a material with a large positive value of $\Delta P_{\mathrm{g,F}}$. The apochromat has a focal length of $f(\lambda_{\mathrm{d}}) = 100\,\mathrm{mm}$ and an aperture of $D = 12.5\,\mathrm{mm}$. The spot diagram in the right inset visualizes that the spot size is only slightly larger than that of the ideal design example presented in the main text (Fig. 6.3). The root mean square spot size obtained from ray-tracing is $d_{\mathrm{RMS}} = 2.2\,\mu\mathrm{m}$. The black circle denotes the Airy disk's diameter of $d_{\mathrm{Airy}} = 5.7\,\mu\mathrm{m}$, which quantifies the diffraction limit. **b** Focal shift $\Delta f(\lambda) = f(\lambda) - f(\lambda_{\mathrm{C}})$ as a function of the wavelength to show that the apochromatization condition $f(\lambda_{\mathrm{g}}) = f(\lambda_{\mathrm{F}}) = f(\lambda_{\mathrm{C}})$ is fulfilled. **c** Partial dispersion diagram visualizing the ν_{d} and $P_{\mathrm{g,F}}$ values of the apochromat's materials. The dotted blue line denotes the normal line for optical glasses

waveguides' material, for example by locally varying the concentration of a dopant or using layered waveguides. From a theoretical standpoint, this is certainly a promising approach but it is questionable if such a device can be fabricated. In conclusion, this analysis shows that a significant amount of work is still required until metalenses can achieve broadband focusing efficiencies over 90 % ($\eta_{\mathrm{PIDE}} > 90\,\%$). Finally, I emphasize again that, even if efficiency achromatization is achieved for metalenses, the maximum achievable efficiencies of such devices would be significantly below 100 % because the ideal continous phase profiles are only discretely sampled [30].

A.9 Apochromat with Dispersion-Engineered Materials

To show that thin layers of materials with large positive values of $\Delta P_{\mathrm{g,F}}$ also allow for the design of a high-performance apochromat, I designed an additional apochromat. As for the designs presented in the main text (Sect. 6.2), I chose a focal length of

$f(\lambda_d) = 100$ mm and an aperture of $D = 12.5$ mm. I also kept the optical glasses K7 and F2 as the first two materials. However, in contrast to the design presented in Sect. 6.2, I assumed a large positive value of $\Delta P_{g,F}$. Specifically, I used a value of $\Delta P_{g,F} = +0.3$, which is equal to the partial dispersion that resulted from the optimization of the apochromat's third material in Sect. 6.2 except for the opposite sign. Furthermore, I kept the material's other properties fixed at $n_d = 1.6$ and $\nu_d = 10$. Other than that I used the same optimization procedure as in the main text.

Figure A.19a depicts the layout of optimized system with $\Delta P_{g,F} = +0.3$. In addition, Fig. A.19b illustrates the system's focal shift (Δf) as a function of the wavelength to show that it is indeed apochromatic. Finally, Fig. A.19c visualizes the materials' locations in the partial dispersion diagram. These figures show that a large positive value of $\Delta P_{g,F}$ can also be used in the thin layer. In fact, the thin layer approximately provides the same amount of refractive power as in the design presented in the main text. This can be seen by comparing the two values for the focal length f_3 in Figs. A.19a and 6.3a. This confirms that the amount of refractive power in the thin layer is mainly determined by $\Delta P_{g,F}$ (see (5.8)). Furthermore, comparing Fig. A.19 to the corresponding figure in the main text (Fig. A.19) shows that the spot size of the system is only slightly increased compared to the system with a negative value of $\Delta P_{g,F}$. It is also evident from the figure that this is because the conventional two lenses need to provide a slightly higher amount of refractive power, that is, their individual focal lengths f_1 and f_2 are slightly shorter. It is well known that this increases the amount of spherical aberration [22]. However, this only has a minor impact on the system's performance and the system still significantly outperforms the one that is composed of conventional optical glasses (Fig. 6.4). Therefore, this design example illustrates that thin layers of materials with anomalous values of $P_{g,F}$ generally allow for correcting chromatic aberrations in broadband optical systems to a high degree. As discussed in the main text, such materials could possibly completely replace so-called special glasses and hence be a key technology for the next generation of optical systems. However, future research is essential to evaluate precisely how much dispersion can be introduced into such materials while still maintaining a sufficiently high transparency.

A.10 Optical Designs for Smartphone Cameras—Performance

In this section, I provide a full overview over the performance of the two designs for smartphone cameras that I discussed in Sect. 6.3. Both designs were optimized for a high imaging performance at three object distances of $l_{obj} = \infty$, $l_{obj} = 1000$ mm, and $l_{obj} = 500$ mm. Therefore, I here present the modulation transfer function (MTF) at these object distances as a function of the spatial frequency. Accordingly, Fig. A.20 depicts the MTF for the benchmark design that was extracted from an Apple patent [31]. Analogously, Fig. A.21 presents the corresponding data for the hybrid design

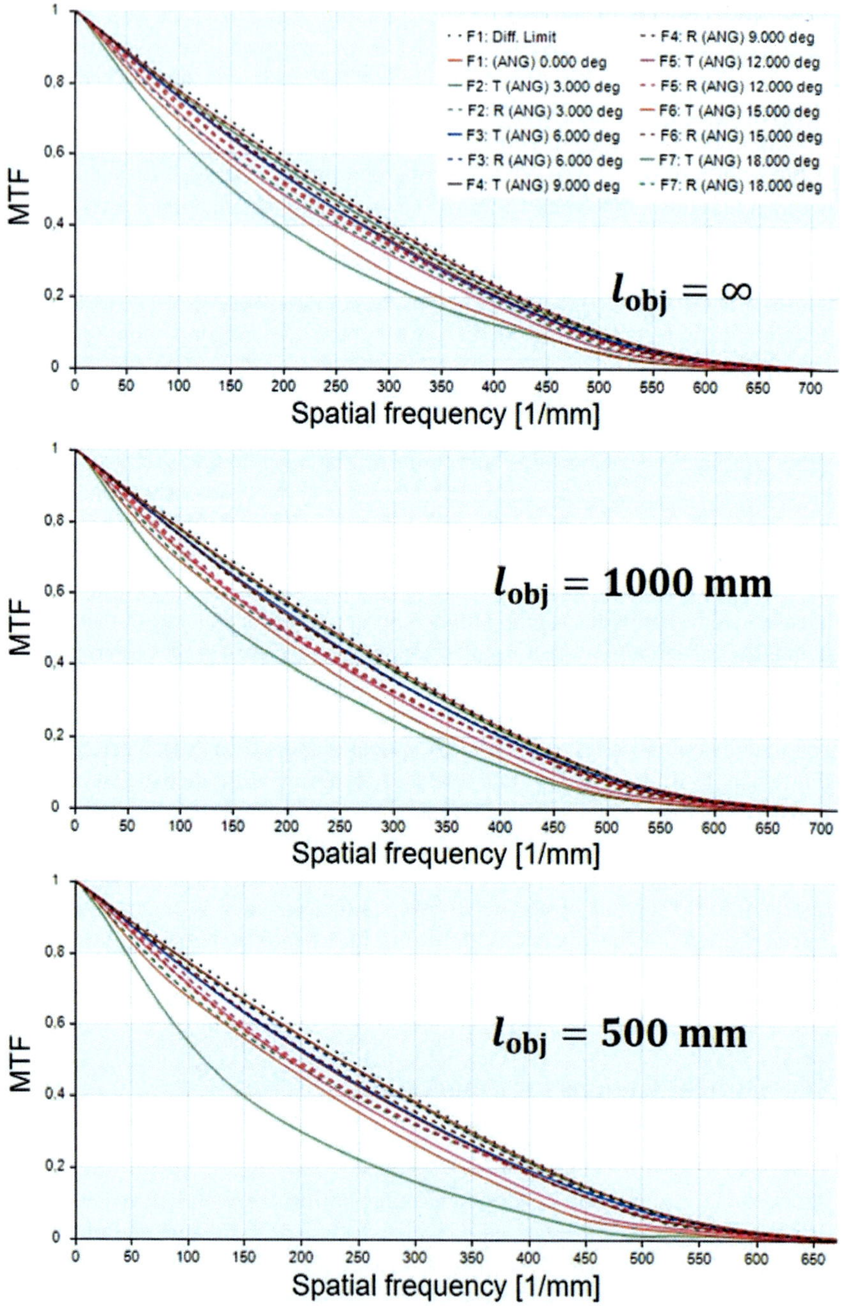

Fig. A.20 Modulation transfer function (MTF) of the benchmark design for a telephoto smartphone camera at different object distances (l_{obj}) over the spatial frequency. This benchmark design was extracted from an Apple patent [31]. All specifications are given in Sect. 6.3

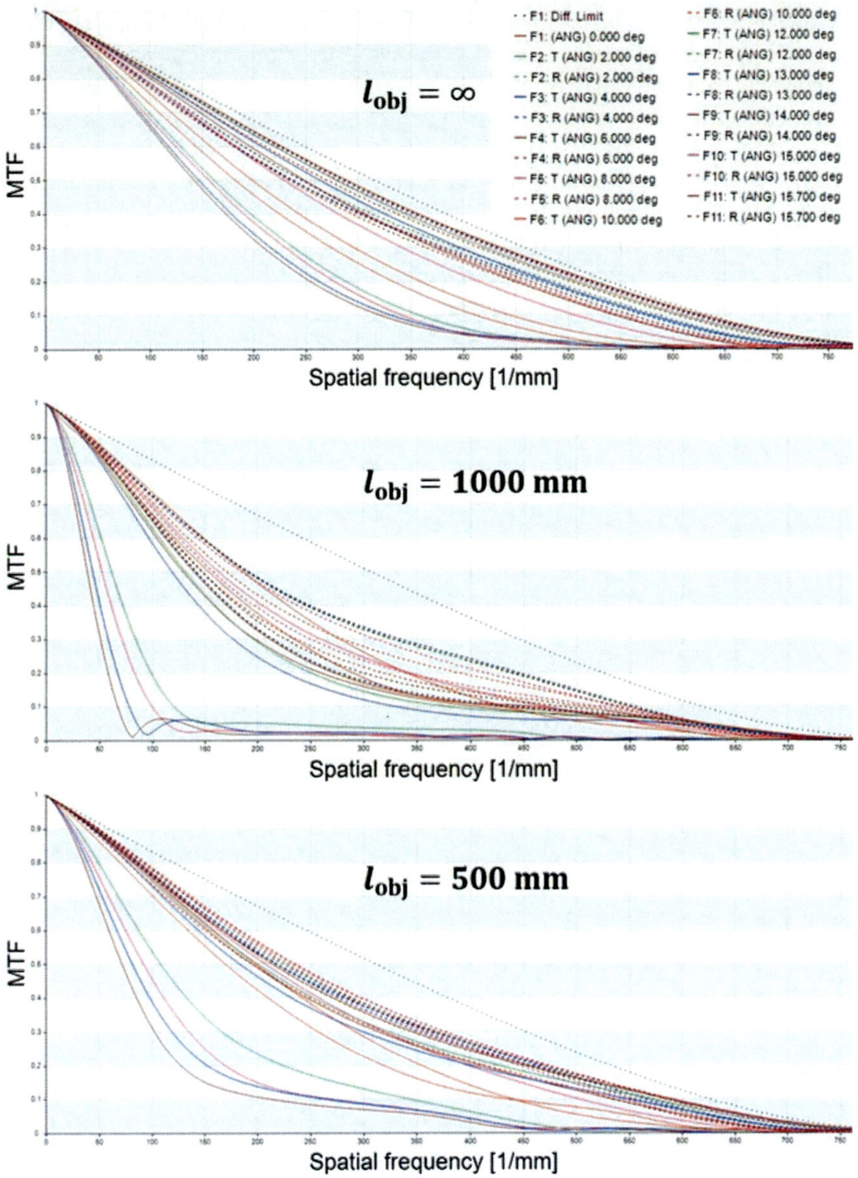

Fig. A.21 Modulation transfer function (MTF) of the hybrid design for a telephoto smartphone camera at different object distances (l_{obj}) over the spatial frequency. This design includes two diffractive lenses. All specifications are given in Sect. 6.3

that includes two diffractive lenses (DLs). All specifications of both designs are given in Sect. 6.3. The comparison of Figs. A.20 and A.21 shows that, at $l_{obj} = \infty$, both designs achieve approximately the same performance (see also Sect. 6.3). Only at the shorter object distances of $l_{obj} = 1000$ mm and $l_{obj} = 500$ mm, the performance of the hybrid design is slightly impaired. This is because we pushed this design's focal length and aperture to the limit at which a high performance was still maintained for the object distance of infinity. In fact, the performance at shorter object distances could be significantly improved by increasing the minimal object distance included in the optimization. Since modern smartphones generally include several camera modules, it therefore appears to be the most promising approach to cover the shortest object distances with other modules.

References

1. D. Werdehausen, X.G. Santiago, S. Burger, I. Staude, T. Pertsch, C. Rockstuhl, M. Decker, Modelling optical materials at the single scatterer level—the transition from homogeneous to heterogeneous materials. Adv. Theory Simul **3**, 2000192 (2020)
2. D. Werdehausen, I. Staude, S. Burger, J. Petschulat, T. Scharf, T. Pertsch, M. Decker, Design rules for customizable optical materials based on nanocomposites. Opt. Mater. Exp. **8**(11), 3456 (2018)
3. C.F. Bohren, Applicability of effective-medium theories to problems of scattering and absorption by nonhomogeneous atmospheric particles. J. Atmospheric Sci. **43**(5), 468–475 (1986)
4. D.R. Smith, S. Schultz, P. Markoš, C.M. Soukoulis, Determination of effective permittivity and permeability of metamaterials from reflection and transmission coefficients. Phys. Rev. B **65**(19) (2002)
5. C. Rockstuhl, T. Zentgraf, H. Guo, N. Liu, C. Etrich, I. Loa, K. Syassen, J. Kuhl, F. Lederer, H. Giessen, Resonances of split-ring resonator metamaterials in the near infrared. Appl. Phys. B **84**(1–2), 219–227 (2006)
6. D. Werdehausen, S. Burger, I. Staude, T. Pertsch, M. Decker, General design formalism for highly efficient flat optics for broadband applications. Opt. Exp. **28**(5), 6452–6468 (2020)
7. B.H. Kleemann, M. Seesselberg, J. Ruoff, Design concepts for broadband high-efficiency DOEs. J. Eur. Opt. Soc. Rapid Publ. **3** (2008)
8. T. Nakai, H. Ogawa, Research on multi-layer diffractive optical elements and their application to camera lenses, in *Diffractive Optics and Micro-Optics* (Optical Society of America, 2002), DMA2
9. T. Nakai, Diffractive Optical Element. Patent US6587272 (1999)
10. T. Nakai, Diffraction Optical Element. Patent US6829093 (2004)
11. T. Nakai, H. Ukuda, M. Nakabayashi, Diffraction Optical Element. Patent US20050116083 (2005)
12. A.J. Glass, K.J. Weible, A. Schilling, H.P. Herzig, D.R. Lobb, J.W. Goodman, M. Chang, A.H. Guenther, T. Asakura, Achromatization of the diffraction efficiency of diffractive optical elements, in Proceeding SPIE 3749, 18th Congress of the International Commission for Optics, vol. 3749 (1999), pp. 378–379
13. D. Werdehausen, S. Burger, I. Staude, T. Pertsch, M. Decker, Dispersion-engineered nanocomposites enable achromatic diffractive optical elements. Opt. **6**(8), 1031 (2019)
14. M. Seesselberg, J. Ruoff, B.H. Kleemann, Diffractive optical element for colour sensor has multiple successive curvatures structure at right angles to extension direction. Patent DE102006007432 (2007)
15. W.T. Chen, A.Y. Zhu, V. Sanjeev, M. Khorasaninejad, Z. Shi, E. Lee, F. Capasso, A broadband achromatic metalens for focusing and imaging in the visible. Nat. Nanotechnol. **13**(3), 220–226 (2018)

D. Werdehausen, *Nanocomposites as Next-Generation Optical Materials*, Springer Series in Materials Science 316, https://doi.org/10.1007/978-3-030-75684-0

16. W.T. Chen, A.Y. Zhu, J. Sisler, Z. Bharwani, F. Capasso, A broadband achromatic polarization-insensitive metalens consisting of anisotropic nanostructures. Nat. Commun. **10**(1), 355 (2019)
17. M. Khorasaninejad, Z. Shi, A.Y. Zhu, W.T. Chen, V. Sanjeev, A. Zaidi, F. Capasso, Achromatic metalens over 60 nm bandwidth in the visible and metalens with reverse chromatic dispersion. Nano Lett. **17**(3), 1819–1824 (2017)
18. S. Wang et al., A broadband achromatic metalens in the visible. Nat. Nanotechnol. **13**(3), 227–232 (2018)
19. S. Banerji, M. Meem, A. Majumder, F.G. Vasquez, B. Sensale-Rodriguez, R. Menon, Imaging with flat optics: metalenses or diffractive lenses? Optica **6**(6), 805 (2019)
20. F. Aieta, M.A. Kats, P. Genevet, F. Capasso, Multiwavelength achromatic metasurfaces by dispersive phase compensation. Science **347**(6228), 1342 (2015)
21. S. Schmidt, S. Thiele, A. Herkommer, A. Tunnermann, H. Gross, Rotationally symmetric formulation of the wave propagation method-application to the straylight analysis of diffractive lenses. Opt. Lett. **42**(8), 1612–1615 (2017)
22. H. Gross, W. Singer, M. Totzeck, F. Blechinger, B. Achtner, *Handbook of Optical Systems*, vol. 1 (Wiley-VCH, Berlin, 2005)
23. D. Werdehausen, S. Burger, I. Staude, T. Pertsch, M. Decker, Flat optics in high numerical aperture broadband imaging systems. J. Opt. **22**(6), 065607 (2020)
24. P. Lalanne, P. Chavel, Metalenses at visible wavelengths: past, present, perspectives. Laser Photon. Rev. **11**(3), 1600295 (2017)
25. M. Khorasaninejad, W.T. Chen, R.C. Devlin, J. Oh, A.Y. Zhu, F. Capasso, Metalenses at visible wavelengths: diffraction-limited focusing and subwavelength resolution imaging. Science **352**(6290), 1190–4 (2016)
26. P. Lalanne, S. Astilean, P. Chavel, E. Cambril, H. Launois, Blazed binary subwavelength gratings with efficiencies larger than those of conventional échelette gratings. Opt. Lett. **23**(14), 1081 (1998)
27. M. Khorasaninejad, A.Y. Zhu, C. Roques-Carmes, W.T. Chen, J. Oh, I. Mishra, R.C. Devlin, F. Capasso, Polarization-insensitive Metalenses at visible wavelengths. Nano Lett. **16**(11), 7229–7234 (2016)
28. U. Wiedemann, K. Karapetyan, C. Dan, D. Pritzkau, W. Alt, S. Irsen, D. Meschede, Measurement of submicrometre diameters of tapered optical fibres using harmonic generation. Opt. Exp. **18**(8), 7693–704 (2010)
29. P. Lalanne, Waveguiding in blazed-binary diffractive elements. J. Opt. Soc. Am. A **16**(10), 2517 (1999)
30. P. Lalanne, S. Astilean, P. Chavel, E. Cambril, H. Launois, Design and fabrication of blazed binary diffractive elements with sampling periods smaller than the structural cutoff. J. Opt. Soc. Am. A **16**(5), 1143–1156 (1999)
31. R.I. Mercado, Small form factor telephoto camera. Patent US2015/0116569 (2015)

Index

D. Werdehausen, *Nanocomposites as Next-Generation Optical Materials*, Springer Series in Materials Science 316, https://doi.org/10.1007/978-3-030-75684-0

J

K

L

M

N

O

P

Q

R